HALF DOME

THE HISTORY
OF YOSEMITE'S
ICONIC MOUNTAIN

HALF DOME

THE HISTORY OF YOSEMITE'S ICONIC MOUNTAIN

JOE REIDHEAD

WITH ILLUSTRATIONS BY
SUSAN GREENLEAF

HALF DOME: THE HISTORY OF YOSEMITE'S ICONIC MOUNTAIN. This edition © 2019 by Joe Reidhead. All rights reserved. No part of this book may be reproduced in any form or by any means without permission from the publisher or author except in the case of brief quotations for use in critical articles and reviews.

Susan Greenleaf created the illustrated maps of the eastern end of Yosemite Valley, the Half Dome Cables, the Regular Northwest Face, and the South Face.

Beryl Knauth created the illustration of a ragged looking climber leading an aid pitch.

All photos are licensed under CC-BY-SA 2.0 or CC-BY-SA 3.0, with the exception of the Galen Rowell, Jerry Gallwas, and Yosemite Archive photos.

ISBN 978-1-940777-79-5

Reidhead & Company Publishers
www.reidheadpublishers.com
Bishop, California

This book is set in Minion Pro, designed by Robert Slimbach for Adobe Systems. It was inspired by the elegant and highly readable typefaces of the late Renaissance.

Dedication

To the bears, mountain lions, deer, ring-tailed cats, ravens, squirrels, jays, salamanders, and rattlesnakes. May they avoid GPS collars and radio trackers, cheesy corn puffs and hamburger patties, and selfies. For the Valley is theirs.

Table of Contents

Introduction		1
1	Geological History	3
2	Ecological History	19
3	Native American History	37
4	European Sightings and Naming a Mountain	43
5	First Attempts to Scale the Dome	49
6	First Ascents	57
7	The Cables	71
8	First Technical Rock Climb	81
9	The Regular Northwest Face	91
10	The South Face	103
11	Managing a Mountain	119
12	A Dangerous Mountain	131
13	Half Dome Today	145
Acknowledgments		153
Bibliography		155
Index		157
The Author		160

by Susan Greenleaf

At the head of the valley, now clearly revealed, stands the Half Dome, the loftiest, most sublime and the most beautiful of all the rocks that guard this glorious temple.

JOHN MUIR

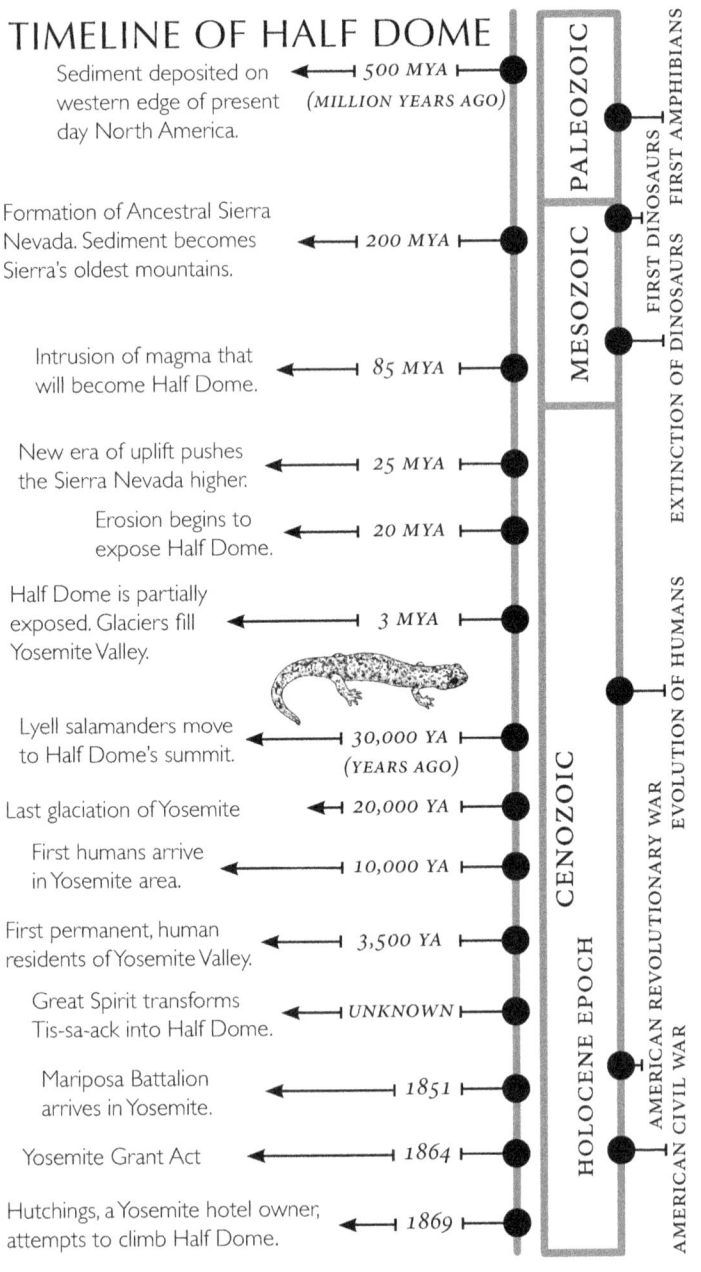

TIMELINE OF HALF DOME

Sediment deposited on western edge of present day North America.	500 MYA (MILLION YEARS AGO)	PALEOZOIC — FIRST AMPHIBIANS
Formation of Ancestral Sierra Nevada. Sediment becomes Sierra's oldest mountains.	200 MYA	FIRST DINOSAURS
Intrusion of magma that will become Half Dome.	85 MYA	MESOZOIC — EXTINCTION OF DINOSAURS
New era of uplift pushes the Sierra Nevada higher.	25 MYA	
Erosion begins to expose Half Dome.	20 MYA	
Half Dome is partially exposed. Glaciers fill Yosemite Valley.	3 MYA	EVOLUTION OF HUMANS
Lyell salamanders move to Half Dome's summit.	30,000 YA (YEARS AGO)	CENOZOIC
Last glaciation of Yosemite	20,000 YA	
First humans arrive in Yosemite area.	10,000 YA	AMERICAN REVOLUTIONARY WAR
First permanent, human residents of Yosemite Valley.	3,500 YA	HOLOCENE EPOCH
Great Spirit transforms Tis-sa-ack into Half Dome.	UNKNOWN	
Mariposa Battalion arrives in Yosemite.	1851	AMERICAN CIVIL WAR
Yosemite Grant Act	1864	
Hutchings, a Yosemite hotel owner, attempts to climb Half Dome.	1869	

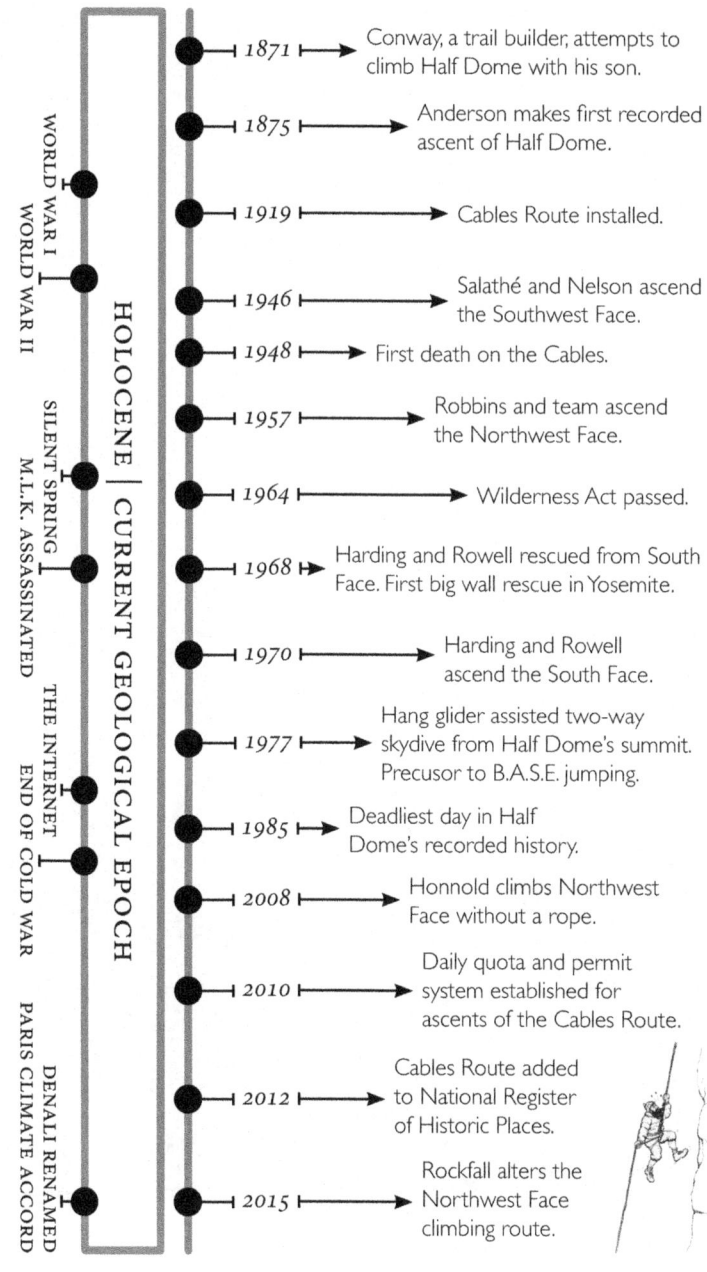

Photo by Vicente Villamón.

Introduction

For several years, I had the fortune to live in the shadow of Half Dome. There, I observed its many moods through every season. I watched lightning pummel the summit and avalanches cascade down the granite slabs. On clear evenings, shadows chased the last golden rays of sunlight up Half Dome's face, and I enjoyed the show from one of the Valley's peaceful meadows. I also saw individuals from around the world create unique and beautiful connections with this spectacular mountain. For me, the sight of Half Dome always reminded me that I was home. For many others, this granite monolith is a symbol of strength, freedom, wilderness, and perseverance against the impossible. Countless hikers and climbers view their ascents to the summit as emblematic of greater challenges that they face in their lives outside of the mountains. As I witnessed the experiences of the mountain and the individuals who came to witness it, I was inspired to tell Half Dome's story.

Often we hear the stories of an individual's ascent of a mountain. We learn of the trials faced by climbers and of their physical, emotional, and spiritual struggles to reach the summit. Less often do we hear the story of a mountain. While mountains may "speak" to our hearts and minds, they do not speak our language. They have always been silent, unmoved guardians of a rarefied landscape where humans may visit but never remain. If a mountain could tell its life story, what would it say?

The story starts with the geological history—Half Dome's birth and formation. It then explores the unique ecology of the summit of Half Dome. Following that are the Native American stories that speak of its spiritual creation. Readers will learn about Half Dome's first encounters with Europeans. I have dedicated ample space to some of the most significant climbs on the Dome. Throughout the book, I explore the history of the mountain as a public space. No mountain is without tragic tales, and those too are in this book.

The hope of *Half Dome: The History of Yosemite's Iconic Mountain* is to create a stronger connection between this land and the people drawn to it. When we understand someone's (or something's) story—the events that have shaped their identity in this world—we tend to have more compassion for them. Why would we need compassion for a block of stone, so grand and old to be almost beyond our comprehension? The natural landscapes of this world have few defenses against our eagerness to build civilizations and tame the wilds. It is increasingly difficult to find wild spaces on this planet. Our human presence dominates the landscape. We have grown distant from the land, causing us to lose compassion for the earth that has given us life. In a way, we have lost ourselves. Yet there is hope in the wilderness that remains. We can still experience these sacred spaces. We can give ourselves to them, and receive so much in return, for nature is the best medicine.

1

GEOLOGICAL HISTORY

IN THE BEGINNING

The formation of Half Dome took place in the womb of the earth about 85 million years ago. The family history of Half Dome, though, goes back in time much further to the birth of the Sierra Nevada mountain range.

500 million years ago, a sea existed along the western edge of North America. The process of erosion deposited vast amounts of sediment along this coast for hundreds of millions of years. This sediment would one day become the building blocks of the oldest peaks in the Sierra Nevada. But how did sediment at the bottom of an ancient sea become the mountain range we see today?

BORN OF FIRE

If we fast forward to about 200 million years ago, the Farallon tectonic plate underneath what is now the Pacific Ocean pushed under western North America. The geologic process of one tectonic plate pushing underneath another plate is called subduction. In the case of the Sierra Nevada, subduction had several effects.

What happened to the other half?

Half Dome's southwest side from Washburn Point. Photo courtesy of the Yosemite National Park Archives.

First, it generated a lot of heat and pressure, which compressed the sediment from 500 million years ago into the metamorphic rocks that now form some of the Sierra's peaks. Metamorphic rocks are born when a pre-existing rock is subjected to extreme heat and pressure, which brings about chemical and physical changes.

As the tectonic plate underneath the ocean pushed under the North American plate, the heat from this friction also caused magma deep beneath the surface of the earth. The magma moved upward through the earth's crust and erupted at the surface, forming a chain of volcanoes along what is now the Sierra Nevada range. This ancient range of volcanoes was similar to the modern Cascade Range in northern California, Oregon, and Washington.

Not all the magma generated by the friction of subduction became volcanoes. Between 200 and 80 million years ago, magma forced its way up and through the more brittle overlying rock, but some of it failed to breach the surface of the earth. These intrusions of magma amassed into giant blobs of molten rock and slowly cooled underground. Cooled and hardened magma is called igneous rock. A large mass of igneous rock is called a pluton. How common are plutons? Within Yosemite alone, there are close to 100 plutons, and our protagonist is one of them.

The Half Dome pluton lies within a region where the dominant granite type is called Half Dome Granodiorite. This granodiorite is about 85 million years old, making it Yosemite Valley's youngest igneous rock. It is also the most common rock in the Valley. Upon examination of Half Dome's granite, curious hikers will notice crystals made of various minerals including biotite, hornblende, and feldspars. These crystals formed as the magma cooled. The large and dark hornblende crystals are of notable beauty.

At this point in the timeline, a distinct mountain range known as the Ancestral Sierra Nevada had formed. Its summits reached up to 15,000 feet. The contemporary Sierra Nevada did not yet exist. Millions of years of erosion and uplift were required to shape the mountains that many people enjoy today.

By 65 million years ago, many of the old volcanoes had lost their breath of fire. Erosion started to expose the underlying masses of plutons, collectively known as a batholith. The Ancestral Sierra Nevada continued to erode faster than it grew. The range transformed into a chain of high rolling hills with none of the fascinating geological features of today's range.

NEW LIFE FOR AN OLD RANGE

About 25 million years ago, things got particularly exciting. Another tectonic plate subducted underneath the worn down mountain range. The subduction of this plate injected new life into the area, and the Sierra Nevada began to grow out of the earth's crust. More precisely, a block of the crust broke free and uplifted. Today, this block gradually climbs out of the west and drops off precipitously in the east where it broke off from the rest of the crust. The Sierra Nevada continues to grow today at a rate of about 1-2 centimeters a year.[1]

From 20 million to 5 million years ago, a new batch of volcanoes erupted throughout the range as more magma found its way to the surface. This volcanic period is evident in various features throughout the Sierra, including the basalt columns of Devil's Postpile and the hot springs of the Long Valley Caldera near the ski town of Mammoth Lakes.

Erosion continued to cut away at the mountains. The higher the mountains rose, the faster the water flowed. Swift rivers and streams carved out steep, deep, and narrow canyons.

The pluton that is Half Dome emerged from the ground as erosion removed the soil around it. The Dome's sheer Northwest Face that towers above the Valley also began to take shape, though by a different process entirely.

GROWING PAINS

Countless joints, or cracks, run parallel to Half Dome's face. These cracks suffer the strain of ever-changing seasons. The freeze-thaw cycles of the colder months pry the cracks apart. The intense heat of the summer swells the rock, putting more stress on weak joints. Over time, flakes of granite

---

[1] Research suggests that during the drought of 2011-2015, the mountains lost so much water that the weight displacement caused the range to rise almost 1 inch! When snow returned to the range in the following years, the mountains shrank half an inch. This data highlights the importance of mountains as water storage and the massive effect of climate change on mountain ranges. In the case of the Sierra Nevada, this range is the source for much of California's water.

peel away from the face like the layer of an onion and fall off, often gouging a path of destruction in the ground below. This process of peeling away from the main body of stone is called exfoliation.

On the ground below the face, the exfoliation debris accumulated over millions of years into piles called talus. Periodically, the talus slid into the Valley below. Today, fresh examples of talus fields are visible on the Valley floor near Mirror Lake.

Visitors to Yosemite might notice the countless cracks zigzagging across Half Dome's face. For rock climbers, cracks like these are the Sirens' song that brings them to Yosemite. When the magma that is Half Dome cooled miles underground, it was under immense pressure from the overlying rock. This pressure released as Half Dome emerged from the ground. The rock expanded without the weight of the earth bearing down upon it any longer. The stresses of expansion in this massive monolith increased until they surpassed the strength of the granite, causing the outer edges of the rock to split into several layers of shells, much like the layers of an onion. The face of Half Dome continued its emergence into the world.

THE ICE AGE

Beginning about two to three million years ago, many of Yosemite Valley's famous features were visible, but lacked the bold definition we see today. The Valley had a deep "V" shape, like the Merced River Canyon to the west. The broad, forest floor and meadows that we know today did not exist.

The Valley and the entire Sierra Nevada went through several ice ages that brought huge glaciers to the region. Geologists have found four distinct glacial periods that occurred in the Sierra: the Sherwin, the Tahoe, the Tenaya, and, lastly, the Tioga.

During these distinct ice ages, the glaciers that poured into the Valley accelerated the process of erosion. Extending to maximum depths of almost 4,000 feet, the glaciers scoured the bottom of the Valley and transformed it into the wide "U" shaped valley that we know.

Here we arrive at a common question: Did a glacier carve Half Dome into the surreal feature it has become?

Interestingly, the geological evidence suggests that glaciers were not a significant player in the formation of Half Dome's cleft face. Instead, the face of Half Dome likely formed via exfoliation that occurred between glaciations. During the earliest and biggest glaciation, the Sherwin, the glacier that dominated Yosemite Valley nearly reached the top of Half Dome, coming within 500 feet of the summit. The last glaciation to enter the Valley, the Tioga, started about 28,000 years ago and never touched the bottom of Half Dome's Northwest Face.

The role the glaciers played in Half Dome's formation was to remove the massive piles of exfoliated rock away from Half Dome's base and out of the Valley. Additionally, the glaciers may have plucked away rock flakes hanging from the face, thus accelerating the process of exfoliation. The glaciers pulverized this rocky debris in their cold depths until they spit it out at their terminus. The roaring glacial-fed rivers then swept the debris downstream. The rivers swept this debris, along with

Above, the slow process of erosion carved out the rolling landscape. Half Dome, a rounded bump, is just noticible at the top center.

Below, after millions of years erosion exposed the granite monoliths hidden beneath the surface and Half Dome's shape became more distinct. Half Dome's face took shape as process of exfoliation unfolded.

Above, glaciers filled Yosemite Valley and the High Sierra beyond. Only the top 500 feet of Half Dome extended above the glacial maximum. Half Dome's face continued to exfoliate above the glacier. The glacier helped exfoliation by plucking off rock flakes below the ice's surface.

Below, the glacier in the Valley retreated, revealing the broad "U" shaped valley that it carved out and polished granite below the glacial maximum.

the older metamorphic rocks, towards California's now-fertile Central Valley.

Over the millennia, exfoliation transformed the face from a series of small steps and broken cliffs into a sheer wall.

HALF A DOME

Thus far, a primary question has gone unanswered: Is Half Dome half a dome?

The answer is no. From the Valley floor, Half Dome appears as a sphere chopped in half, but this view is deceptive. For a more accurate perspective, one must visit the nearby Glacier Point. From there, Half Dome appears as a thin fin of rock and it is evident that the South Face is almost as sheer as the more famous Northwest Face. This view allows visitors to understand why geologists estimate that 80% of Half Dome remains intact.

The geological story of Half Dome did not end when the last glacier receded from the Valley. Exfoliation continues today. For climbers, this fact became evident in July of 2015 when a rock flake 200 feet tall and 100 feet wide fell off of one of Yosemite's most famous rock climbs, the Regular Northwest Face of Half Dome. While Half Dome's geological history took place over hundreds of millions of years, these processes can still create cataclysmic events.

While this unique geologic story has made Yosemite famous today, there are many creatures who have learned to thrive amidst the rocks. Let's now explore these living characters of Half Dome.

Watercolor by Marguerite Zorach, 1920.

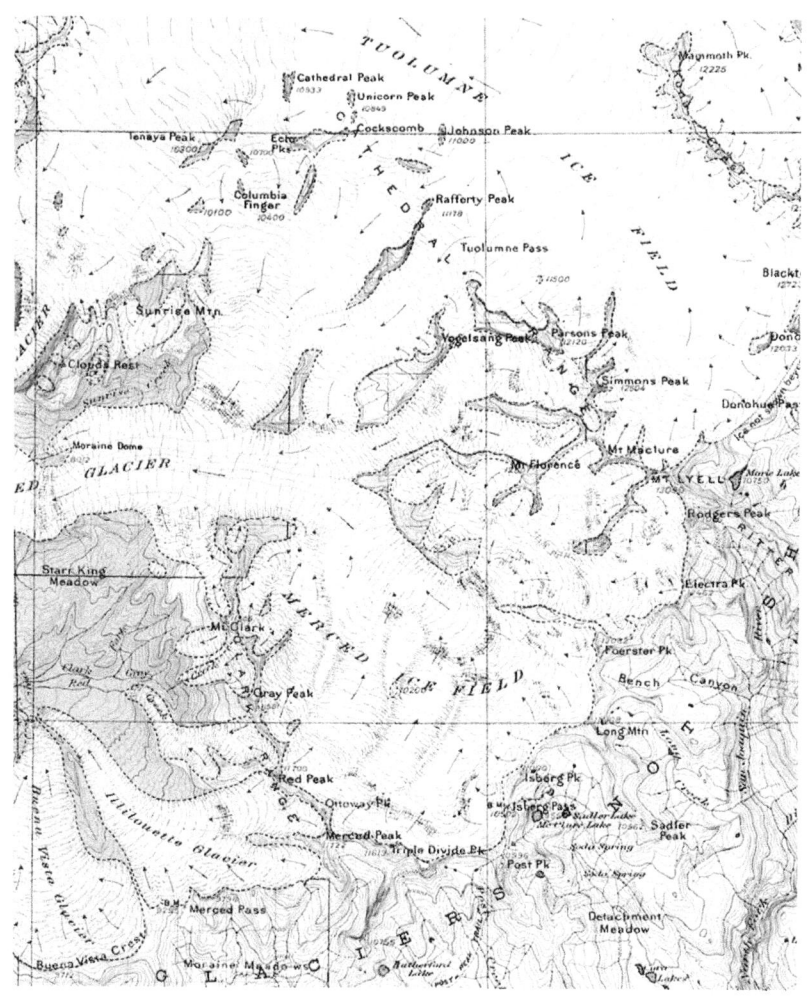

LABELED MAP OF YOSEMITE VALLEY AND BEYOND

RS	Rockslides	HD	Half Dome
RF	Ribbon Fall	M	Mount Maclure
EC	El Capitan	L	Mount Lyell
TB	Three Brothers	F	Mount Florence
EP	Eagle Peak	CC	Cascade Cliffs
YF	Yosemite Falls	BP	Bunnell Point
IC	Indian Canyon Creek	LY	Little Yosemite Valley
R	Royal Arches	LC	Liberty Cap
W	Washington Column	B	Mount Broderick
TC	Tenaya Canyon	SD	Sentinel Dome
ML	Mirror Lake	G	Glacier Point
ND	North Dome	SR	Sentinel Rock
BD	Basket Dome	T	Taft Point
MW	Mount Watkins	CS	Cathedral Spires
E	Echo Peaks	CR	Cathedral Rocks
C	Clouds Rest	BV	Bridalveil Fall
SM	Sunrise Mountain	LT	Leaning Tower
Q	Quarter Domes	DP	Dewey Point
TC	Tenaya Canyon	MR	Merced River

Half Dome's Northwest Face and the Merced River in autumn. The Southwest Shoulder is the right skyline of Half Dome. Courtesy of the Yosemite National Park Archives.

2

Ecological History

Most hikers might not think of the summit of Half Dome as a unique and vibrant ecosystem, but it is precisely that, as are many other summits in California's Sierra Nevada mountain range.

THE LYELL SALAMANDERS

The most famous and endearing of Half Dome's year-round residents are the Mount Lyell salamanders, *Hydromantes platycephalus*. The color of dark chocolate, the Lyell salamander is a member of the *Plethodontid* family—the lungless salamanders. The Lyell salamanders breathe through their skin and mouth tissues, and this requires a humid environment that will keep these delicate tissues moist They have found this wet environment under the rocks and snowfields of the Sierra Nevada.

These salamanders have lived on Sierra summits for almost 30,000 years, before the last of the glaciers receded from Yosemite Valley. They were unknown until 1915 when two were caught accidentally in a researcher's mouse trap near Mount Lyell.[2] While scientists have found these salamanders

---

2   Mount Lyell is the highest peak in Yosemite National Park with a summit elevation of 13,120 feet.

as low as 4,000 feet in elevation, most live above 7,000 feet, with Half Dome's congress of salamanders living at a summit elevation of 8,839 feet.

The Lyell salamander has a comfortable home atop Half Dome; living beneath the seasonal summit snowfield, under the summit rocks, and deep inside the cracks and fissures that shoot through the Dome.

Curiously, the Lyell salamanders' nearest relatives within the *Hydromantes* genus are found far away in Italy and southern France. How it came to be that this genus of salamanders is found only in the Sierra and southern Europe is a mystery.

In the winter, the salamanders retreat into warm underground habitats to avoid the freezing temperatures that they cannot tolerate. Once spring has thoroughly taken hold of the Sierra, the salamanders stir from their slumber.

As nocturnal creatures, they emerge under cover of night to feast on countless insects and spiders. In feeding, the thin-skinned salamanders establish themselves as true predators. Hidden behind a gentle face is the salamander's food capturing weapon. A long, sticky tongue darts out of the salamander's mouth and ensnares the night's feast.

These salamanders have one other important weapon to use during the hunt. Sticky, webbed feet allow the salamanders to explore and hunt on cliff faces. In this vertical world, they can find plenty of insects to feed on and deep, moisture-laden cracks to call home.

Half Dome's Lyell salamanders do not have many natural predators thanks to the steep granite faces that drop off from all sides of the summit plateau. The mice and marmots who

Courtesy of the United States Forest Service.

share this space are not regular predators of the salamanders. Instead, the salamanders' primary enemy lurks in the skies above Half Dome. Ravens and other birds patrol the skies to scoop up the five inch long salamanders when they find the opportunity.

The salamanders do have a few self-defense techniques, however. First among these are toxic skin secretions and a sacrificial tail. When a salamander senses a threat, it will flatten its body to the ground, arch its tail above its body, and release toxins through its skin. The tail secretes many of the toxins, which are slightly poisonous and have a foul taste to discourage predators. The tail also serves as a decoy; if grabbed, it detaches from the salamander's body and allows the creature to run away. If a salamander loses its tail, the appendage will regenerate within a few weeks. In addition to these defenses, the salamander is a nimble escape artist during the heat of a battle. It will coil its body into a ball and quickly roll downhill or into the crevices of Half Dome's stacked boulders. As it bounces along, it will often spring out of its ball, flipping through the air to change direction and confuse a predator, and then return to a coiled ball to flee the scene.

Yosemite Valley in the summer is known for being hot and dry, characteristic of its Mediteranean-like climate. During these dry spells, the salamanders retreat once again to their underground dwellings to slip into a state of dormancy called aestivation. As summer transitions to fall, thunderstorms return moisture to Half Dome's summit, allowing the salamanders to reemerge until autumn's freezing temperatures and snowfall force them deep below ground until the spring.

While all of this may sound like an idyllic life—living atop Half Dome, eating and sleeping the days away with few natural predators—the salamanders here face a vicious threat to their existence: humans. Some years ago, when camping was still allowed on the summit, campers built rock shelters to create more comfortable sleeping spots and block out the mountain winds. Campers gathered the rocks from the summit plateau, thus disturbing the moist underground homes of the salamanders. For this reason, camping is no longer allowed on the summit; these small, sensitive creatures whose population is quite small and scattered need our respect and protection.

Today, the salamanders still face threats from hikers who unknowingly use the salamanders' shelters to construct rock cairns—towers of rocks stacked one atop the other for ostensible artistic or navigational purpose. Some of the cairns are over six feet tall and marvels of engineering, but a cairn of any size is a destruction of habitat in this sensitive ecosystem. Park Rangers and volunteers break down the cairns as often as possible, but the tide of self-proclaimed artists and amateur trail builders continues to endanger this threatened species. Instead of leaving a mark on nature, perhaps we all would be better served allowing nature's beautiful vistas to imprint upon us.

MARMOTS

If the Lyell salamander is the most endearing of creatures atop Half Dome, then the yellow-bellied marmot must be the most notorious. These stout, furry rodents regularly steal hikers' unattended food and chew on the sweat-soaked straps of backpacks.

The yellow-bellied marmot, *Marmota flaviventris*, is one of the larger members of the squirrel family. The marmot has a coat of thick, brown fur, except for the namesake yellow streak adorning its chest like a knight's breastplate.

The chubby marmots living on Half Dome can weigh up to 11 pounds, surpassing the size of the average house cat. They are diurnal, meaning active during the day, and this explains the oft-reported sightings by hikers. While the marmots are omnivorous, their diet is mostly limited to grass, leaves, and flowers. Insects and the occasional bird egg supplement this fiber-rich diet. Given a choice, the marmots most likely avoid eating the mildly toxic and foul tasting Mount Lyell salamanders.

In obtaining water, the marmots are also well-adapted to this barren environment. Their primarily plant-based diet provides them with ample hydration. Of course, Half Dome's human visitors are not so efficiently adapted and should carry up to 3 liters of water for the hike, or risk dehydration.

Lucky hikers will witness a running marmot. Why are

A yellow-bellied marmot near
the summit of Mount Dana.
Photo by Inklein.

they so lucky? Marmots are not graceful runners. A marmot run is more of a quick waddle—their fat stomachs swaying from side to side. Their short legs struggle to move faster while their disproportionately large bodies still lumber along. The marmots are better adapted to their underground burrows, where they spend up to 80% of their lives.

Fortunately for Half Dome's marmots, there is little need to run. Foxes and coyotes, their usual predators in the Sierra Nevada, are not present on the summit plateau. Eagles, however, do ply the skies above, and the marmots must not grow complacent atop this seeming refuge. When a predator does appear in the air, any marmot who witnesses the threat will let out a sharp whistle to warn the other members of its colony, hence the nickname "whistle-pig."

Marmots can live for up to fifteen years. They live in colonies of up to twenty individuals. In the spring, the marmots emerge from hibernation and are ready to mate. For the male, this means finding up to four mates in a season, though he will hibernate with only one of them. Some 30 days later the females give birth to a litter that averages four baby marmots, but she can have up to eight pups.

The typical day in a marmot's life is a day of extended rest. Like most humans, they wake up, poop, and groom themselves. Then they lounge in the sun until mid-morning, when it's time to find breakfast. After breakfast, they might spend more time basking in the sun or hanging out in their burrow. Dinner takes place in the late afternoon before they retire to the burrow for the night. Observant hikers may see members of a colony playing and grooming each other.

In September, with cooling temperatures and advancing winter storms, Half Dome's marmots fatten up and prepare for their winter hibernation. For up to eight months the marmots will slumber in their cozy burrows beneath the summit's boulders.

Hibernation may appear to be a quiet, restful time in a marmot's life. However, many marmots die in hibernation during the long, dark hours of the winter. Another common source of marmot mortality is the plague. The plague is common among rodents in the High Sierra, where it occurs naturally. This is a good reason never to touch these cute but wild creatures!

Marmots prefer to live in high elevation meadows near the edges of talus fields or forests. Hikers might wonder how marmots came to exist on the summit of Half Dome. Did the marmots' ancestors walk across a long-gone glacier? Or perhaps a female and a male escaped the talons of a golden eagle and landed on the summit?

The answer is none of these. The marmots found their way to the summit via the same route as today's hikers. They ascended the low angle granite slabs where the Cables Route is bolted to the rock. If this fact is surprising, consider that the Yosemite Black Bear can ascend technical rock climbing routes that most humans cannot climb with ropes and harnesses! Yosemite's mammals have evolved with a unique penchant for climbing.

RAVENS

The common raven, *Corvus corax*, is common to Yosemite Valley, the Sierra Nevada, North America, and many other parts of the world. However, their habitat wasn't always as large as it is today. Ravens are highly intelligent, surpassing apes, and even humans, in some cognitive tests. The success of the raven is directly tied to the success of humanity, for they have followed our roads and developments, using them as sources of food, water, and shelter, thus expanding their vast habitat.

Ravens are large birds with wingspans stretching up to 25 inches. Their stout bodies and jet-black coats shimmering with iridescence give the ravens an almost ominous feel. And if that does not cause a second-glance, then surely their eerie clicks, crackles, and mimics will cause many a visitor to give them a wide berth.

If ever there were an ideal home for a raven, it must be Yosemite Valley. The massive cliff faces and towering pine trees are a giant playground for the acrobatic ravens who roll, somersault, and dive through the Valley's famous vistas. Visitors standing on Half Dome can often look over the edge to see a raven or two gliding through the air below or catching a warm thermal to ascend effortlessly above the Valley's rim.

Ravens typically mate for life. The search for the right partner takes time, sometimes up to three years for eager juveniles ready to mate. Courtship among ravens involves demonstrations of flying, intelligence, and food gathering. After courting, the pair establishes a territory and builds a nest in one of Yosemite's tall trees or on a cliffside.

A raven in Camp 4. Photo by David Iliff.

The diet of ravens is opportunistic. They eat what they can find. Within a week, a raven's food could include: squirrel roadkill on the Yosemite Valley Loop, a mouse on Half Dome's Sub Dome, invasive Himalayan blackberries in the Valley, an abandoned scrap of pizza at the Yosemite Valley Lodge, and spilled trail mix atop Half Dome's "visor". In their opportunism, the ravens are no different from many of Yosemite other famous residents: the black bear, the elusive ring-tailed cat, or the "impoverished" rock climber ascending Half Dome's 2,000-foot-tall Regular Northwest Face.

For many cultures, ravens are a sign of evil, even death—they are to be viewed with suspicion and avoided. What has the raven done to deserve this bad rap? They are incredibly intelligent creatures. They quickly make friends of many humans. Indeed, they can even recognize a unique human face. Perhaps ravens frighten us because their behavior resembles ours so closely. While some cultures steer clear of ravens, some contemporary human residents of Yosemite believe the ravens there carry the spirits of deceased friends and family—a symbol of life after death.

Remember—when you are standing on the summit of Half Dome, and *Corvus corax* lands next to you, its rattle call punctuating the mountain wind—you can smile and say hello, but don't let it trick you into surrendering your food, or else you will discover firsthand which species is more intelligent.

PINE TREES

Modern-day hikers know there are no trees on Half Dome's summit, but this was not always the case. The disappearance of Half Dome's pines is another sad tale of humans altering this landscape for the worse.

When a plucky Scot, who we will meet later, made the first ascent of Half Dome in 1875, there were four patches of pines on the summit with a combined total of seven trees. Three species were present then: *Pinus jeffreyi* (Jeffrey), *Pinus contorta* (Lodgepole), and *Pinus albicaulis* (Whitebark). Of the seven trees, only one survives out of sight and safe from hikers. What follows is an arboreal obituary of sorts.

Pinus jeffreyi, the Jeffrey pine, is also known as "gentle Jeffrey" for its smooth pine cones. Upon sighting a Jeffrey, many a mountain rambler will stick their nose into its bark to inhale the butterscotch scent of its resin.

Pinus contorta, the lodgepole pine, is named for its traditional use by Native Americans in the construction of tipis and lodges. Of course, visitors to Yosemite must remember that the Native Americans of Yosemite Valley, the Ahwahneechee, constructed cedar bark houses and not tipis.

Pinus albicaulis, the whitebark pine, is another alpine pine common to Yosemite. Its strong roots cling to summits and exposed ridges in a state of *krummholz*, in which a lifetime of fierce winds and winter storms force the tree to grow into a gnarled and twisted dwarf. The whitebark pine's propagation is dependent on a winged alpine friend to spread its seeds

A weather beaten *Pinus jeffreyi* in a state of krummholz at Olmstead Point in Yosemite National Park. Note Half Dome and Clouds Rest in the background.

through the mountains—the Clark's nutcracker, *Nucifraga columbiana*. The Clark's nutcracker caches the seeds underground for times of scarcity, but when the nutcracker fails to retrieve the seeds, a clump of whitebark pines is often born.

What disaster caused the deaths of the magnificent, weathered, and gnarled trees that once lived atop the Dome?

Humans. In particular, unprepared campers who cut down the trees for firewood. If this tragedy can teach a lesson, it is this: let us be conscious of seemingly small actions in the present, for they may appear more significant in the future.

OTHER FLORA

Half Dome's summit is not so barren a place as some might have you believe. In addition to the mentioned animals and trees, there is an abundance of flora scattered around this island in the sky.

There are alpine *spiraea*—a genus of hardy deciduous shrubs—and there are the delicate wildflowers of *potentilla* (cinquefoils), *erigeron* (daisies), *eriogonum* (wild buckwheat), *penstemon* (beardtongues), and *solidago* (goldenrods). A variety of grasses can be spotted by the observant eye as well.

The most seen but least recognized residents of Half Dome are the dark streaks running down its steep face. Some people may know these streaks as the sorrowful tears of Tis-sa-ack, the female protagonist of a legendary Ahwahneechee tale. We will explore her story in the next chapter. As one might guess, the streaks are not hardened tears. They are communities of rock lichens.

Communities of dark lichens create the tears of Tis-sa-ack on the Northwest Face.

A lichen is a composite organism. It is the symbiotic combination of a fungus and a photosynthetic partner (algae or cyanobacteria) growing together. The lichens that form Tissa-ack's tears are crustose, meaning they grow tight and flat against the rock. Several types of lichen live in the streaks, including *Lecidea atrobrunnea* and *Dimelaena thysanota*.

Estimates suggest that lichens cover some 6-8% of the earth's land. They are among the oldest living organisms on the planet, possibly reaching ages of 10,000 years old! As with other long-living organisms, they grow extremely slow. Along the Mist Trail section of the hike to Half Dome, visitors who wish to leave their names etched into the stone scrub off defenseless mosses and lichens. It takes the native lichens and mosses a long time to recover from this eco-graffiti, if ever.

THE HUMAN IMPACT

In the account of his ascent of Half Dome, John Muir wrote:

> "For my part I should prefer leaving it in pure wildness, though, after all, no great damage could be done by tramping over it. The surface would be strewn with tin cans and bottles, but the winter gales would blow the rubbish away."

These words were written more than 100 years ago, and our attitudes towards litter have certainly changed since then. Though Muir was a great defender of Yosemite, I must disagree with the acceptable standards of his day; trampling over Half

Dome's summit does cause considerable harm. It is a majestic and wonderful place to visit, but let us all tread softly through this fragile ecosystem.

What about those tin cans and bottles? They must blow somewhere.

Each autumn, a special event takes place in Yosemite. The event is the annual Yosemite Facelift, a weeklong cleanup effort organized by Yosemite's rock climbing community and the non-profit Yosemite Climbing Association. The mission is to clean up Yosemite after the busy summer season. As a part of this effort, the National Park Service's climbing rangers and volunteers rappel off of Half Dome's cables and remove trash that falls onto the rock ledges below. I once participated in this and we packed out some eighty pounds of garbage in one afternoon from the Half Dome area.

For conservation to work, everyone must participate. Pack out everything that you pack in. Do no harm. Give the landscape and its creatures the respect you would ask to be given to you, or more.

Reconstructed traditional cedar bark dwellings called u-ma-cha. Photo courtesy of the Yosemite National Park Archives.

3

NATIVE AMERICAN HISTORY

Before the world knew of Yosemite's towering granite walls and majestic waterfalls, a small group of Native Americans knew it as home.

The first humans arrived in the area as long as 7,000 to 10,000 years ago. Around the same time, the last of the mighty glaciers of the Tioga Glaciation retreated deep into the mountains.

The earliest people did not inhabit Yosemite Valley year-round. It would have been a challenging place to live, not long after the powerful glaciers and the rivers fed by their meltwater had scoured the Valley. Instead, the first permanent residents of the Valley appeared approximately 3,500 years ago. This marked the beginning of a unique culture that eventually became the Southern Sierra Miwok, whose descendants still live in the foothills and often gather in Yosemite to continue their heritage.

The earliest residents hunted with spears and atlatls—a spear-throwing device. A hunter using an atlatl was a highly skilled individual and essential member of the group. A skillfully wielded atlatl provided meat, which meant protein and vital nutrients for the tribe. But to throw a spear incorrectly was to put one's family and extended family at risk, and potentially lose a prized obsidian spear point.

The tribe hunted a variety of game, including deer, bears, and smaller mammals. They gathered wild edibles that included berries, pine nuts, mushrooms, and insects. The staple of their diet was the acorn of *Quercus kelloggii*, the black oak tree. They stored the acorns in a granary called a chuck-ah. When they were ready to eat the acorns, they used pounding stones to pulverize the seeds into flour. These stones can still be seen in Yosemite Valley today. Once the acorns were turned into flour, the flour was then soaked to remove tannins. The talented cooks then turned the mush into porridge or tasty flatbread.

The tribes of the Valley often started controlled fires. They were wise stewards of this land and knew the importance of fire in maintaining a healthy and balanced ecosystem. The brushfires also had the benefit of killing undergrowth that competed with the black oaks. In turn, there were more oaks to produce the all-important acorn. We must be thankful to Yosemite's first people whenever we sit in one of the Valley's meadows today. They created and maintained many of these peaceful spaces.

They lived in cedar bark dwellings similar in shape to a tipi, but more permanent. These homes were called u-ma-cha. The u-ma-cha were waterproof, warm, repelled insects, and could house an entire family.

The older generations passed on their traditions and creation stories to the younger generations. Many special ceremonies took place in large roundhouses and sweat lodges. Over the millennia, the groups living in the Valley changed. With these changes, cultures merged and evolved.

The last group of Miwok to live in Yosemite Valley called it "Ahwahnee." Ahwahnee translates to "place of the gaping mouth." Being residents of Ahwahnee, they called themselves the "Ahwahneechee," meaning "dwellers in Ahwahnee."

The Ahwahneechee traded with their neighbors, including the Paiutes and Monos. From the Paiutes of the eastern Sierra, they obtained the precious obsidian stone necessary for crafting spear and arrow points. From the Mono, they received the delicious ka-cha-vee, an insect pupa that tasted like shrimp, and caterpillars that live in the Jeffrey Pines of the Sierra high country.

Unfortunately, the Ahwahneechee's time in the Valley was not destined to last. The discovery of gold in the western foothills brought a massive influx of European prospectors to their doorstep. The miners' search for the yellow rock was feverish and consuming. In the process, they damaged the ecosystems that sustained the local tribes, including the Ahwahneechee. Thousands of Miwok who lived in the western foothills of the Sierra and the Central Valley either died from starvation or directly at the hands of the miners. When the Miwok tried to resist, the newcomers responded with brutality.

THE GOLD RUSH AND THE TRAGEDY OF CALIFORNIA'S NATIVE AMERICANS

1851 saw the formation of the Mariposa Battalion, led by James Savage. Savage was a miner and store owner who had trading posts and gold mines in the foothills, including a trading post downriver from Yosemite and another on the Fresno

River. After a group of Native Americans raided his store in 1850, he rallied for the formation of a retaliatory militia. With approval from the state of California, the militia set off to find Yosemite Valley and capture the Ahwahneechee.

The militia had never contacted the Ahwahneechee before. When they launched their assault on the Native American villages in the Valley, they failed to ask what name the Native Americans had given this place. Instead, the militia members adopted a name that other Miwok tribes used for this band, "Yohhe'meti" or "Yos-s-e'meti." Yosemite is translated as "those who kill." Their neighbors gave the name Yosemite to the Ahwahneechee because they were known to have outcasts and renegades from other tribes amongst their numbers.

The militia captured the Ahwahneechee and burned their villages over the course of several days. They then marched Chief Tenaya, the leader of the Ahwahneechee, and his people to a reservation on the Fresno River in the Central Valley, far from their ancient home.

To add insult to injury, the Mariposa Battalion named the Valley "Yosemite," and they mistranslated it as "grizzly bear." They also appropriated the Chief's name to rename several features near where they captured the tribe: Tenaya Lake, Tenaya Peak, and Tenaya Canyon. Lafayette Bunnell, a member of the Battalion, recounted:

> "I called [Tenaya] up to us, and told him that we had given his name to the lake and river. At first, he seemed unable to comprehend our purpose, and pointing to the group of glistening peaks,

near the head of the lake, said: 'It already has a name, we call it Py-we-ack.' Upon my telling him that we had named it [Tenaya], because it was upon the shores of the lake that we had found his people, who would never return to it to live, his countenance fell and he at once left our group and joined his own family circle. His countenance as he left us indicated that he thought the naming of the lake no equivalent for the loss of his territory."

HER NAME WAS TIS-SA-ACK

The story of the Ahwahneechee, the California tribes, and all of the Native Americans is a tragedy. While the Ahwahneechee are no longer residents of Yosemite, they did leave us with a few creation stories. One of these is the legend of how Half Dome came to be. The Ahwahneechee, however, knew the mountain as "Tis-sa-ack."

This story takes place long before the Ahwahneechee came to live in the Valley. A woman and her husband left the river plains of the Central Valley to visit their relatives across the high mountains to the east. It would be a long journey. The woman's name was Tis-sa-ack, and the man was Nangas. The trip was tiring as they slogged up the dry and hot foothills. Finally, they arrived in the Valley exhausted and thirsty.

At the head of the Valley, they came to a broad section of the river known as Ah-wei-ya. Today we call it Mirror Lake.

Tis-sa-ack placed her baby in its basket and drank from the lake. She was so thirsty that she filled her cup again and again. No amount of water could quench her thirst. When Nangas arrived a short moment later to drink from the lake, not a drop remained.

Anger consumed Nangas. He rushed to his wife and started to beat her. Tis-sa-ack tried to flee from the onslaught, but she could not escape his blows. Tears filled her eyes. Finally, in great pain, she faced her husband and threw her basket at him.

This violence angered the Great Spirit. Nangas and Tis-sa-ack had disturbed the peace of the Valley. In retaliation, the Great Spirit turned them to stone. Their abandoned baby became the Royal Arches, and it stands above today's Majestic Hotel (originally named the Ahwahnee Hotel). Nangas became Washington Column and North Dome. The thrown basket became Basket Dome, which sits aside North Dome. Tis-sa-ack became Half Dome. The dark streaks running down its Northwest Face are her tears of pain. And thus Tis-sa-ack is often translated to "face of young woman stained with tears."

Visitors to Yosemite Valley can learn more about the story of Yosemite's Native Americans by visiting the Indian Museum, where members of the Southern Sierra Miwok (some of whom are descendants of the Ahwahneechee) and National Park Service park rangers keep the stories alive.

The legend of Tis-sa-ack raises important questions. Why do we call it Half Dome and not Tis-sa-ack? Who gets to decide such things as the name of a mountain on a map? And how does the name of a mountain impact the native cultures that have cherished it and the collective memories and stories that endure?

4

EUROPEAN SIGHTINGS AND NAMING A MOUNTAIN

The naming of Half Dome shared the same fate as the naming of many mountains and landscapes in the United States. Often, there was already a Native American name for distinguishing features. Then European explorers and settlers changed those names to words more familiar to them. In some cases, surveyors and politicians further changed the names given by the earlier Europeans.

The first English language name bestowed upon the mountain was most likely "Rock of Ages." Two bear hunters by the names of Reamer and Abrams gave the mountain this Christian name in 1849. Abrams' journal entry reads:

> "Reamer and I saw a grizzly bear['s] tracks and went out to hunt him down getting lost in the mountains and not returning until the following evening, found our way to camp over an Indian trail that led past a valley enclosed by stupendous cliffs rising perhaps 3,000 feet from their base and which gave us cause for wonder. Not far off a waterfall dropped from a cliff below three jagged peaks into the valley while farther beyond a rounded mountain stood, the valley wide of

43

which looked as though it had been sliced with a knife as one would slice a loaf of bread and which Reamer and I called the Rock of Ages."

Abrams' estimate of the height of the Valley's walls is the first accurate guess at their height. Future inaccurate guesses put El Capitan at a half to a quarter of that size.

The present name came courtesy of the Mariposa Battalion, the same group who came to Yosemite Valley to remove the Native Americans. On March 27, 1851, while scouting away from camp, a Private Champion Spencer became lost deep in thought about the mountain. His Sergeant, Alexander Cameron, remembered:

> "Spencer looked a good long while at that split mountain, and called it a 'half dome.' I concluded he might name it what he liked, if he would leave it and go to camp for I was getting tired and hungry and said so."

Private Spencer's deep pondering is not an inspiring origin story for the naming of Half Dome. It is no equal to Tis-sa-ack and her story, but Half Dome is the name that stuck.

The Mariposa Battalion had several non-white members. These were the Native American scouts they had recruited from other tribes to guide them into the Valley and assist in tracking the Ahwahneechee. Some of these scouts had bad blood with the Ahwahneechee from previous conflicts and were willing to help the Battalion. Some of the scouts knew Half Dome as "The Sentinel." In the present day, The Sentinel is now assigned to the jagged granite spire towering over the Four Mile Trail, across the Merced River from the Yosemite Valley Lodge and Camp 4.

Later, Half Dome was briefly called "South Dome." The intent behind this name was likely to create geographical harmony by having a North Dome and a South Dome. While this name did not endure, it was used by John Muir in several of his writings, including the description of his ascent of Half Dome in November 1875.

Perhaps Muir preferred South Dome. Regardless, he was a decade behind the consensus of his fellow pioneers. In 1865, Clarence King and James Gardiner, two surveyors working for the California Geological Survey, affixed the name "Half Dome" to the mountain on the official maps. Josiah Whitney, the chief of the Survey and namesake of California's Mount Whitney, also used "Half Dome" in his 1869 book "The Yosemite Valley." It is also important to note that Whitney made no mention of "South Dome" or "Tis-sa-ack" in the same book.

A MONUMENT TO THE AMERICAN ETHOS

In 1864, President Lincoln signed the Yosemite Grant Act—a pivotal moment in the public lands movement. The Act marked the first time that the United States Government set aside land for protection and public enjoyment. It also set the tone for the creation of Yellowstone in 1872, the first National Park.

The Yosemite Grant took place amidst the horrors of the American Civil War. The Grant reflected a realignment of political policies within the federal government founded in emancipation and with the goal of rewarding Americans for their sacrifices during the Civil War. What better way to highlight the greatest benefit of American citizenship—freedom—than to set aside one of its most majestic landscapes in perpetuity for all people?

At the same time, the landscape photographs of Carlton Watkins and the paintings of Albert Bierstadt reached the East Coast. These photos and paintings depicted the peaceful and inspiring scenes of Yosemite Valley. They were an island of the sublime amidst an ocean of atrocities. They were a reminder of the American ethos of liberty as reflected in its landscape.

During this time, suggestions for Half Dome's name included "Goddess of Liberty" and "Mt. Abraham Lincoln." Both of these names are understandable given the political climate of the time; they reflected the hope for a better future that existed amidst a time of great upheaval.

Mountains have always held a special place in the psyche of humans. Half Dome is a perfect example of this phenomenon. Highlighting this is another of Half Dome's first suggested names, "Spirit of the Valley." To this day Half Dome captures the hearts of many. People travel to Yosemite from all over the world to visit this mountain that stirs powerful feelings within them.

The history of the naming of Half Dome demonstrates the significance we attach to these seemingly inaccessible places. We see them as monuments to the past, atonement for sins committed, signs of strength to contrast the inherent frailty of humanity, and symbols of our ideals.

Liberty Cap and Nevada Fall. Oil painting by Albert Bierstadt, 1873.

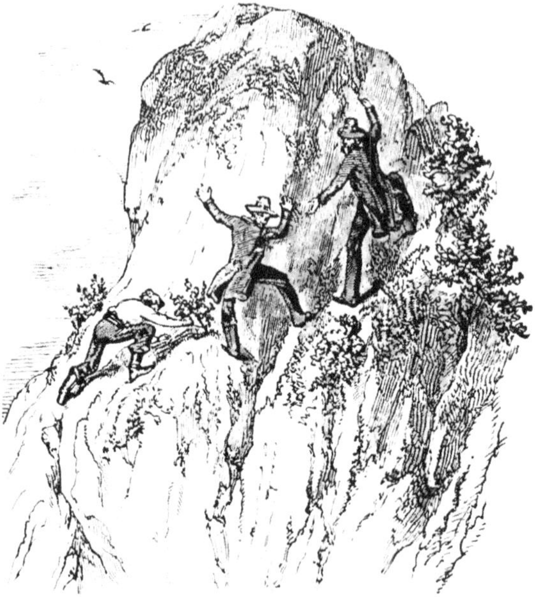

5

FIRST ATTEMPTS TO SCALE THE DOME

With some mountains, stating who was the first to reach the summit can be a tricky and even contentious business. Throughout the Americas, there is archaeological evidence of indigenous ascents before the "First Ascents" recorded by European mountaineers. Indeed, some contemporary mountaineers prefer to use the phrase "First Recorded Ascent."

In the Sierra Nevada, firm archaeological evidence of Native American ascents is thin. Clarence King and fellow mountaineers of Josiah Whitney's California Geological Survey claimed to have found various artifacts on some of the summits they attained in the mid-1800s. An arrowhead was reported to have been found on the summit of Mount Whitney, the highest point in the continental United States.

Considering the above, it is reasonable to ask: Did Native Americans ever ascend to the summit of Tis-sa-ack?

The first European settlers of the Valley to reach the summit never reported finding any evidence of a Native American ascent. Given the technical nature of Half Dome before the installation of ropes and cables, this is not surprising. That said, the Anasazi of the desert Southwest were known to be superb climbers who carved their dwellings into the sides of cliffs.

Would it be a far stretch to think that the Ahwahneechee, who lived in one of the world's greatest rock climbing playgrounds, were also adept at scaling rock faces? We may never know; some things will remain a mystery forever.

A PERILOUS ROUTE

For the aspiring Half Dome climbers of the late 1800s, the easiest route to the summit was obvious: the Northeast Shoulder. Hidden from viewers who are down in the Valley or standing atop Glacier Point, the Northeast Shoulder is the lowest angle route to the summit. However, being the least steep of Half Dome's sides did not imply an easy ascent.

To get to the base of the Northeast Shoulder, potential climbers first had to hike or ride horses 4,000 vertical feet up and almost eight miles into the wilderness. They followed Native American trails and newly constructed hikers' trails past Vernal and Nevada Falls, through Little Yosemite Valley, and then through the steep old-growth pine forest that flanks the southern and southeastern side of the Dome.

The beginning of the Northeast Shoulder rises steeply out of the forest floor. This first exposure of rock is the Sub Dome, and its steep face presented the first challenge to would-be climbers. To this day, many hikers cannot complete this climb. They are turned around by the daunting drop-offs next to the trail. We must remember that the earliest climbers did not have the benefit of this trail.

Once atop the Sub Dome, the granite narrows to a thin ridge which then drops down to a saddle. From this saddle rises the mass of Half Dome. It is 400 vertical feet from here to the summit. The granite slab that must be ascended reaches angles of almost 50°—rock that is steep enough to be considered technical rock climbing even by today's standards. The modern Cables Route did not yet exist, nor had it even been conceived.

AN IMPOSSIBLE MOUNTAIN

So, who was the first human to attempt this daring feat? The written record of first attempts to scale Half Dome begins in 1869.

James Hutchings was an entrepreneurial carpenter, gold miner, and journalist who first came to Yosemite in 1855 to see for himself the grand sight of a thousand-foot-tall waterfall that the Mariposa Battalion had reported during their removal of the Ahwahneechee. Over the following years, Hutchings made several trips to Yosemite. He wrote articles for magazines and newspapers describing the grandeur of the Valley. In 1864, he and his wife acquired a small, rustic hotel in the Valley and helped to usher the tourism boom that continues to this day. They called their hotel Hutchings House and transformed it into a comfortable inn over the following years.

We can imagine some of the conversations that took place at the Hutchings House with their guests. During dinners and aperitifs, the subject of Half Dome's impossible summit was discussed time and again. Of course, any discussion about ascending Half Dome would have included speculation about the fame that would come to the person who made the first ascent.

Perhaps it was these conversations that propelled Hutchings towards his attempt at being the first to climb Half Dome in 1869. Hutchings and two others followed a precipitous route from the Valley floor to the saddle between Sub Dome and Half Dome. Here is Hutchings' account:

> "In the summer of 1869 three of us set out for the purpose of climbing [Half Dome], taking the "Indian escape trail" north of Grizzly Peak. There was absolutely no trail whatsoever, as we had to walk on narrow ledges, and hold on with our feet as well as hands, trusting our lives to bushes and jutting points of rock. In some places where the ledges of rock were high, their tops had to be

reached by long broken branches of trees, which the Indians used to climb; and, after they were up, cut off the possibility of pursuit from enemies, by pulling up these primitive ladders after them. Not a drop of water could we find. A snow bank increased rather than diminished our terrible thirst. Finally, after many hair-breadth escapes, and not a little fatigue, we reached the top of the lower dome, or eastern shoulder, and were then within four hundred and sixty feet, vertically, of realizing our ambitious hopes. To our dismay, as well as disappointment, we found a great smooth mountain before us, standing at an angle of about 40°, its surface overlaid and overlapped, so to speak, with vast circular granite shingles, about eighteen inches in thickness. There was not a place to set a secure foot upon, or a point that we could clutch with our fingers. The very first sight put every hope to flight of reaching its exalted summit by the means at our command; and, deeming it a simple impossibility, "we surrendered at discretion," and returned without the realization of our ambitious hopes."

Hutchings was not the only prominent Yosemite chronicler to declare Half Dome an impossible ascent. Josiah Whitney, California's chief geologist, said Half Dome was "perfectly inaccessible, being probably the only one of all the prominent points about the Yosemite which never has been, and never will be, trodden by human foot."

ROPES, BOLTS, AND A CHILD

The next recorded attempt was in September of 1871 by John Conway and his son, Major. Conway made a career as one of Yosemite's most famous trail builders. He cut the Yosemite Falls Trail, the Four Mile Trail, and the Clouds Rest Trail. He was a strong climber. His son, at the mere age of 9, also had a reputation as a climber. They carried a rope and eye-bolts. The father and son drove the bolts into cracks in the rock and secured the rope to the bolts.

At about 300 feet above the saddle between Sub Dome and Half Dome, Major couldn't find any more outcrops to attach the rope to or any placements for the eye-bolts. Conway decided it was too dangerous to continue higher and called his son down. The daring son in the lead thought he could reach the summit, but his father's exhortations prevailed, and they retreated.

John Muir sensationally recounted the Conways' attempt to the San Francisco Bulletin:

> "John Conway, a resident of the valley, has a flock of small boys who climb smooth rocks like lizards, and some two years ago he sent them up the dome with a rope, hoping they might be able to fasten it with spikes driven into fissures, and thus reach the top. They took the rope in tow and succeeded in making it fast two or three hundred feet above the point ordinarily reached, but finding the upper portion of the curve impracticable without

laboriously drilling into the rock, he called down his lizards, thinking himself fortunate in effecting a safe retreat."

If only Muir had lived a hundred more years, he would have seen the potential of such human lizards rock climbing on the Valley's walls.

It took four years for another aspirant to come forth. For this man, "Half Dome fever" became all-consuming.

HALF DOME IN WINTER, YOSEMITE NATIONAL PARK

Gunnar Widforss, 1922.

6

First Ascents

Many people who have heard of Yosemite have heard of John Muir. A Scot by birth, Muir became the most famous voice of Yosemite and an early American proponent of conservation of public lands. However, there is another Scot who will forever be a part of Half Dome's history.

A MOUNTAIN OBSESSION

George Gideon Anderson was born in 1839 in the port town of Montrose on Scotland's eastern coast. He worked as a sailor, but by 1867 he had found his way to the Sierra Nevada foothills during the California Gold Rush.

By 1875, Anderson was in Yosemite Valley, having failed at gold mining. The previous attempts to scale Half Dome were well known to Valley locals. These failures stirred in Anderson an obsession to become the first person to stand on the summit.

Anderson was known in the Valley as a strong and hard worker. Years of physical labor had made him a burly and muscular man. His arms were covered in tattoos, likely reminders of his sailing past. He had a thick beard—a characteristic feature of the mountain men of that day.

Sometime in 1875, Anderson made his way to the saddle between Half Dome and Sub Dome. The Sub Dome was sometimes called the "Camel's Back" and, in the days before a trail was cut into the rock, presented its own exposed rock challenges, but Anderson quickly surmounted this obstacle to reach the top of the Sub Dome. Above him sprawled the final 400 vertical feet of steep granite slab that so far had rejected all previous attempts.

Anderson's first efforts were inventive. He first tried to claw his way up the slab wearing heavy, slick-soled boots. He immediately realized that his boots lacked the friction necessary to climb the slick granite slab. Sliding back down to the saddle, Anderson would have felt the precarious exposure of the situation. Either side of the saddle drops off over thousand-foot cliffs, and the 400-foot high slab of Half Dome's shoulder angles toward one of those cliffs.

Anderson knew that he needed more friction for his feet. He took off his boots and attempted to climb up in his socks. This effort met the same failure. Next, he went barefoot. Many climbers, even today, know that the pads of our feet offer a lot of friction and allow us to grab onto small rock holds with our toes. However, for Anderson, the slab was too steep for this to work.

George Anderson was determined. For his next attempt, he tied rough sackcloth to his feet, but this did not offer the needed friction. Though the bare burlap did not work, Anderson had an idea.

The early rock climbers of Fontainebleau, a forest filled with towering sandstone boulders outside of Paris and near the palace of Versailles, often applied rosin to their mountaineering boots. Rosin is a sticky substance derived from pine pitch, and it helped the climbers stick to the rock. Anderson used a similar technique in his next attempt.

He descended to the pine forest that sits at the base of Sub Dome. There he harvested pitch from the pines and applied it to the sackcloth. Back at the shoulder, he wrapped the sackcloth around his feet once again, with the pitch-covered side facing out. Then he started to climb up the slab. He moved higher and higher on the slab. The pitch worked. It kept his feet from sliding down the granite.

Anderson's sticky shoes did present a new problem. The pitch was so sticky that he had trouble freeing his feet to move higher. As any climber knows, a stuck foot can be fatal. It can cause a climber to lose his balance and fall over. The force required to free each foot could have thrown Anderson over his heels; this would have resulted in a fatal tumble down the slab. After several of these near misses, Anderson retreated once again.

For some, repeated failures can be dejecting. They may sulk away in depression, never to return to the task that defeated them. For others, failure after failure after failure only results in more determination. Anderson ascended into the latter category.

He returned to the Valley with a plan. He bought a drill, a hammer, eye-bolts similar to what the Conways used, wooden pegs, and lots of rope.

Yosemite Valley is a place prone to rumors; both in the past and in the present. Perhaps this is because it is like any small, isolated community. In the case of the first ascent of Half Dome, there was a rumor of a rumor. The talk around the Valley held that the person who first summited Half Dome would have first right to build a hotel either at the saddle or just below Sub Dome. With such incentives—the summit and a hotel—Anderson would have had competition.

After gathering his equipment, he disappeared from the Valley without notice. Of course, then and today, wandering into the mountains alone and without telling anyone is dangerous. After a few days, locals grew concerned with his unannounced absence, and they organized a search party.

Meanwhile, Anderson was busy drilling his way up Half Dome over the course of several days. Hand-drilling bolt holes into granite is a slow and tiring process. One must tap the top of the drill with the hammer while slowly rotating the drill. Anderson had to drill the holes about 3 inches deep. Then he drove in a wooden peg followed by an eye-bolt. The peg served as a wedge to keep the eye-bolt from falling out. He secured his rope to the bolt, pulled himself up, stood on the bolt, and repeated.

In a few spots, Anderson was able to climb along natural cracks or ledges for twenty feet or so before having to place another bolt.

Anderson continued this determined and laborious effort for hundreds of feet. Then, after overcoming several steep sections, including the bare slab that shut down the Conways, George Anderson rounded the shoulder and arrived on the sprawling summit plateau of Half Dome. It was 3 P.M. on October 12, 1875—a crisp fall day in Yosemite.

Anderson drilling a hole on his way to the summit.

Before him, he found a flat summit of about 7 acres, seven clumps of pine trees, and various grasses and shrubs. He might have thought the summit to be an excellent place to camp, with its towering views of the Valley below. He would have hopped along the summit boulders to the high point, 8,839 feet above sea level. He walked over to the precipitous edge of the Northwest Face, the rocky ground just beyond his toes dropping 2,000 feet to the talus below. He was over 4,800 feet above the Valley floor, just a few hundred feet shy of one vertical mile. He walked over to the edge of the Visor, not far from the true summit. The Visor juts out incomprehensibly over the 2,000-foot void, with nothing but air below it. Anderson must have screamed words of victory while marveling at the sights below him from this lofty stance.

Around this time, Anderson's would-be rescuers had started their search. They knew he was fixated on Half Dome, so they started there. They ascended the rough trail that went past Vernal and Nevada Falls. On today's trail, Nevada Fall is about four and a half miles from the summit of Half Dome. And here they ran into Anderson. They asked where he had disappeared to. His response, "Gentlemen, I have been to the top of Half Dome."

Little did Anderson know, his drilling up the shoulder of Half Dome was likely the first instance of true aid climbing in Yosemite. Aid climbing is a specialty within rock climbing where climbers use bolts, pitons, and other pieces of protection to pull themselves up the rock. This is in contrast to free climbing, where climbers only use their hands and feet to pull themselves up. Today, aid climbing is still an important

Anderson standing on the Visor of Half Dome. By S.C. Walker.

aspect of climbing the big walls of Yosemite Valley, such as El Capitan. Climbers come to Yosemite from all over the world to practice and perfect this craft.

SUMMIT TOURISTS AND DREAMS OF RICHES

Anderson left his ropes fixed to the bolts so that he could guide others to the summit. By the end of October, the first guided party, a group of British tourists, made their way to the summit with Anderson's help. A few days later he guided Miss S.L. Dutcher to the summit, the first woman to stand there.

Not long after his ascent, Anderson set himself to prepping the route for future clients. He invested in a softer and stronger rope, tied knots in it for handholds, and fixed it to a strong bolt near the summit and the eye-bolts he placed on his first ascent.

Anderson was a dreamer. He knew that most people were incapable of climbing hand-over-hand up his fixed ropes. He wanted to create an easier way for people to attain the summit. Around 1876, he revealed his plan to build a stairway from the Sub Dome saddle to the summit that would replace the ropes. But his dream of ferrying tourists to the summit did not end with a staircase. Eventually, he hoped to install steam-powered tram cars on the side of Half Dome so that tourists could ascend to the top without effort.

Unfortunately for Anderson, he never made much progress on his stairway aside from amassing timber beams. He did, however, build a new trail from the Valley to Vernal Fall. The process of cutting the trail required extensive blasting

Miss Sally Dutcher was the first woman to hike to the top of Half Dome. Courtesy of the Yosemite National Park Archives.

into the cliffs and heavy rock work. The quality of Anderson's trail work exceeded what anyone had expected at the time. Initially, the state of California commissioned this trail. Anderson hoped to extend the trail all the way to Half Dome, but California withdrew from its agreement with Anderson and withheld payment for already completed work. But here Anderson's hardworking ethic persevered, and he continued building the trail with his own money. Parts of this trail are still in use today.

Anderson continued guiding tourists up Half Dome through the late 1870s and early 1880s. Many of these tourists stayed the night at the hotel of Emily and Albert Snow. They initially called their hotel the Alpine House, but later changed the name to Casa Nevada. Located below Nevada Fall, the hotel provided a convenient staging place for summit bids. The hotel was close enough to the waterfall that the spray from the falling water misted over the hotel's porch. From 1870 until 1889, the Snows provided tourists with warm beds, hot meals, fresh pies, and plenty of liquor. Those who succeeded in summiting Half Dome got to put their name in the register along with their summit date. In many ways, Casa Nevada was similar to the backcountry huts that dot the landscape of Europe's Alps and assist its many mountaineers. The Snows eventually abandoned the hotel, and then in 1900 it caught fire and burned to the ground, bringing to a close the romantic early days of Half Dome hiking.

Anderson never found fortune in his efforts to get more people to the summit. One day in the spring of 1884, Anderson needed money and found a job washing the side of a cabin in the Valley. A cold spring storm moved in while Anderson was cleaning the siding, and, despite pleadings from the owner, Anderson continued working through the storm. The cold took to him, and he came down with pneumonia. He died a short time later on May 8th and was buried in Yosemite's Pioneer Cemetery where visitors can find his grave today.

TWO YOUTH IN SEARCH OF AN ADVENTURE

The winter of 1883-1884 was a heavy snow year in the Sierra. The shoulder of Half Dome where Anderson had attached his ropes to the rock is prone to avalanches. Snow and ice collect on the slick granite until their weight becomes too much to adhere to the rock any longer. When the snow slides off, it does so as massive, cohesive slabs that can destroy anything in their path. The avalanches of this long and snowy winter carried away most of the ropes Anderson had put in place.

After Anderson's death, there was no one in the Valley willing to repeat his ascent and reestablish the ropes. Then, two young men showed up in the Valley looking for adventure. One was from Denver and the other from New York City. They arrived at Glacier Point where they met Galen Clark, the Park's first Guardian. As the men marveled at Half Dome, Clark informed them that the ropes were no longer there and none but the most skilled European mountaineers would be able to replace them.

This statement of near impossibility drove the two men to action. They secretly decided they would climb Half Dome and put the ropes back in place. If they failed, no one would ever know, and they would lose no face. If they succeeded, then the glory was theirs.

With two hundred feet of rope, A. Phimister Proctor and Alden Sampson left their camp at Little Yosemite Valley and headed for Half Dome. They set up another camp near the base of the Dome and immediately set to work.

First, they gathered up Anderson's old rope that now lay at the base. They would reuse it on the route. While many of Anderson's iron eye-bolts were still anchored into the rock, the avalanches had ripped out more than a few. Some of these bolts lay on the ground and were attached to the old rope. The missing bolts would create difficulties.

Proctor chose to climb barefoot, which allowed him to wrap his toes over the eye-bolts. Sampson wore hobnailed mountaineering boots, the standard climbing footwear of the day.

To ascend the blank slab, the two men found a unique rhythm. First, Proctor tied a slipknot into the end of their new rope, creating a lasso. He then eyed the next eye-bolt above him and threw the rope toward the bolt, just as he had seen the California ranchers rope their cattle. Once the rope was lassoed around the bolt, one of the men ascended the rope. Then, they fixed Anderson's old rope to the bolt they had just roped.

Proctor's bare feet were best suited to standing on and gripping the eyebolts. It was a precarious stance. He had to balance with only one foot on the bolt. Then, he had to heave the rope with incredible accuracy to the next bolt, which sometimes was thirty feet away.

After several hours of this exposed and tedious process, they arrived at a 100 foot section where every bolt was gone. Proctor couldn't make the 100 foot throw to the next bolt. Doubt swirled in their minds. How would they overcome this stretch?

Proctor's bare feet, sweaty and sore, offered no friction for free climbing up the steep slab. However, Sampson's hobnailed boots proved equal to the task. He was able to use natural cracks and small edges in the granite to climb up the slab. In a letter to Hutchings, Sampson recounted:

> "The sensation was glorious. I did not stake my life upon it, for I was sure I could make it. If I had slipped in the least I should have had a nasty fall of several hundred feet."

Sampson had tied the rope around his waist on the off-chance that it might arrest his fall if he slipped. But, deep down, he knew the force would break the rigid rope. Finally, he came to the most challenging section—a small roof with slick granite above and below. The granite was bare of holds. But within reach was a small shrub, some 8 inches tall. Sampson latched onto it and gently pulled himself over the roof. The bush held firm, and Sampson reached the next bolt. Even among today's rock climbers, such a long and exposed span would cause the heart to flutter.

By the end of the day, they had fixed ropes halfway up the shoulder. They descended back to their camp as night fell. They were exhausted, proud of what they had accomplished so far, and filled with anticipation for the coming day.

The next day the two men ascended their newly fixed ropes and continued up. But soon they came to another section where the next bolt was 100 feet away. Unfortunately, the granite here was too slick to climb. Proctor threw the lasso over and over for two hours. When they were close to giving up, he made a remarkable throw and caught the bolt. They immediately shouted in joy for they knew they would make it to the summit that day.

After Proctor and Sampson's ascent in 1884, Half Dome saw many attempts and successful ascents over the coming decades. Every so often, climbers had to replace sections of rope and drive new bolts into the rock. Over the decades, the ropes rotted and the bolts rusted. At least one party pulled out a bolt after lassoing it. It is a wonder that no one ever died on the route.

Bold, tenacious, bootstrap mountaineers had pioneered the route up Half Dome. But did they have any foresight of the numbers who would soon follow in their steps? The year 1919 would bring a dramatic change to this lonely and difficult summit.

7

The Cables

While George Anderson may have had the first plan to build a stairway to the summit, he was not the last. Anderson's successor in this effort was Matthew Hall McAllister.

McAllister was a successful businessman from San Francisco who had a deep love for the outdoors, especially the mountains of California. He was an active member of the Sierra Club in the days when the Club still had a strong regional focus on mountain climbing in the Sierra.

A MOUNTAIN FOR ALL

In the early years of the twentieth century, McAllister began to think of ways to build a safe route to Half Dome's summit. But why do this? Why create a route to this remote and challenging summit?

One of his great passions was to encourage others to explore the outdoors and help them to discover sublime moments in nature that would inspire them toward the conservation of these wild places. Those moments could be skiing through an old growth forest, catching a rainbow trout, or reaching the summit of a majestic peak. In Half Dome, with its heavenly 360-degree views, he may have seen a quick path to inspire new generations of conservationists.

McAllister agonized over how to finance, construct, and manage a human-made route up the mountain. Finally, with great fear, he pitched his plan to the Sierra Club.

First, to make the deal as sweet as possible, he would pay for the materials and labor. He would also oversee the work, which would take place as an official Sierra Club project. The Sierra Club, in turn, would pitch this project as a donation to the Park. All the Park had to do was provide in-kind contributions of pack animals to bring the laborers and materials to the job site and tools to perform the work. Further, the plan had the support of David Curry, the owner of the famous Camp Curry located below Glacier Point, who hoped to increase the number of visitors to his hotel.

The directors of the Sierra Club accepted the plan with enthusiasm, and the National Park Service gave its permission as well. Construction started in the Spring of 1919.

But the big question remained: How would McAllister construct a route up the steep, hard, and slick granite shoulder of Half Dome? And could it survive the avalanches that had destroyed George Anderson's ropes?

THE IRON WAY

McAllister's plan was simple. There would be nothing extravagant—no gondolas or 800-foot long staircases. He would install a permanent version of what Anderson had done.

Mules and laborers brought supplies to the base of the Sub Dome. Here, McAllister and his crew started work on what we know today as the Half Dome Trail. First, they had to build a

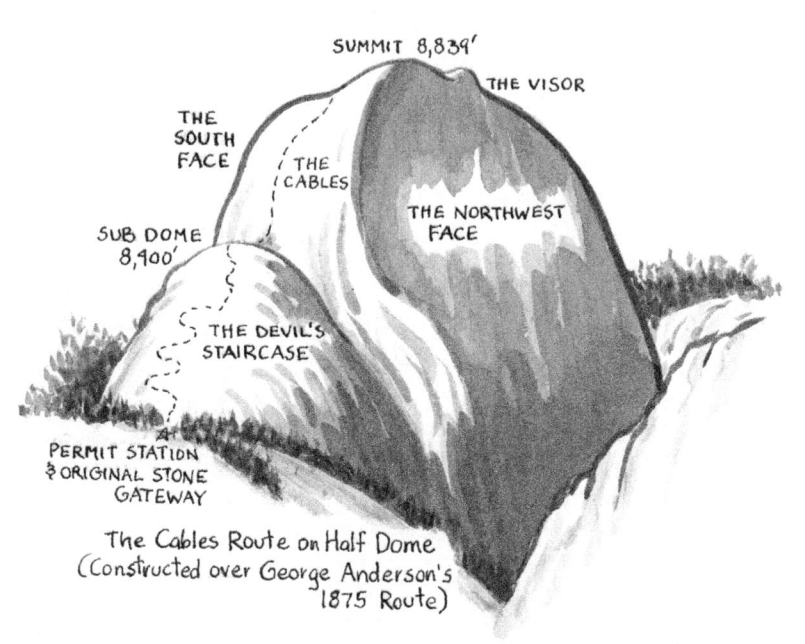

The Cables Route on Half Dome
(Constructed over George Anderson's 1875 Route)

by Susan Greenleaf

better route up the Sub Dome. They blasted away at the rock to make steps. They built traditional dry stone walls to support the trail. What was once a perilous journey in itself, the path up the Sub Dome was now accessible to any hiker. These steps became known to some as the Devil's Staircase.

Higher up at the Sub Dome-Half Dome saddle, they assembled their materials and tools and then set to work. Instead of ropes, the new route used thick, steel cables. There were two sets of cables, placed thirty inches apart. Every ten feet, the workers drilled a large hole into the rock for each cable. They put a metal post, called a stanchion, into the hole. Laborers then placed the cable in a groove in the top of the stanchion. A cap was then screwed on to hold the cable in place. At the bottom of each set of stanchions, a beam was secured to create a resting point. The two cables were anchored to the rock every one hundred feet using giant eye-bolts that were similar to what Anderson had used over fifty years prior.

This path of parallel cables formed a safe walkway. Hikers would climb between the cables, holding on to them for support. When someone needed to rest, all they had to do was stop at a stanchion where a beam created an artificial ledge. For a brief period, there were even harnesses that hikers could use to attach themselves to the cables in case they fell. This new route was undoubtedly in a new spirit compared to the loose bolts and rotten ropes of the previous decades. With these cables, nearly anyone in good physical condition could get up the mountain.

McAllister and his workers completed the project in July of 1919 to much fanfare.

The Yosemite trail crew erecting the Cables for the 1939 hiking season. Photo courtesy of the Yosemite National Park Archives.

The Director of the National Park Service stated in a report:

> "A very useful contribution to the park was made this year [1919] by the donation, through the Sierra Club, of a protection for the trail to the top of Half Dome. ... The old arrangement was dangerous and unsafe. ... The new cable was installed early in July, and it was used by climbers who appreciated keenly the opportunity of seeing the wonderful view from the top of Half Dome, with its sheer drop of practically 5,000 feet to the valley below."

The laborers erected a gateway at the base of the Sub Dome, the foundations of which still stand today. And they placed a plaque near the bottom of the cables, memorializing the project. It read:

Erected
1919
Under Auspices of the
Sierra Club
To Remember
Captain George Anderson
Who First Ascended this Dome in
1875

A couple of years later, National Geographic magazine published photos of the new Cables Route and Half Dome. The Sierra Club Bulletin wrote:

> "The cable stairway up Half Dome, donated to the park by Mr. McAllister, proved very satisfactory, and enabled thousands to reach the summit of the Dome, which heretofore had been a very hazardous undertaking."

For the first time, any person of reasonable fitness could attain Half Dome's summit. The cables were a permanent installation designed to pass the test of time, but can any human-made path truly compare with the permanence of Half Dome?

HALF DOME'S WINTER FURY

During the winter months, snow accumulates on the slick slabs of Half Dome's shoulders. When the snow load reaches a critical mass, it breaks free of the rock and slides down the slabs in destructive, and sometimes deadly, avalanches. McAllister had thought about this and developed a clever plan to prevent his cables from being demolished.

While the cables were affixed to the rock permanently, the stanchions were removable. Each fall, all the Park had to do was unscrew the caps on the stanchions, lift the cable out, and remove the stanchions from their holes. Then the stanchions and wood beams would be stored in a safe place on the Sub Dome. The cables would now rest flat against the granite and thus allow avalanches to slide over them. In the spring, after the ice and snow clinging to Half Dome's shoulder had melted, the process would be reversed.

Unfortunately, the Park Service failed to take down the stanchions during the winter of 1919-1920; perhaps the

A group ascending the Cables Route in 1924 using rope tethers and carabiners to secure them to the cables. Photo courtesy of the Yosemite National Park Archives and Mrs. Wallace.

rangers were busy elsewhere or perhaps they were ignorant of the power of snow. Avalanches destroyed about one hundred feet of stanchions. The cables were still intact, a stroke of luck, and the Park was able to make repairs that summer.

For some reason, the Park did not learn a lesson from this lapse. The cables again suffered avalanche damage in the winter of 1921-1922. One of them even broke free.

The cables had to be repaired several times over the coming decades. Significant repairs and cable replacements took place in 1934 and again in 1984.

The original route that George Anderson had established was dangerous and not for the meek. Very few people made the complete ascent in the years 1875-1919. With the cables in place, hikers flooded in over the coming decades, presenting new challenges to the Park. Separately, the Golden Age of rock climbing in Yosemite was fast approaching. On this tide of change, a few brave souls sought out challenges that far surpassed the engineering feats of Anderson and McAllister. For these climbers, only the steepest of Half Dome's faces were worthy trophies.

Boating Through Yosemite with Half Dome in the Distance. Oil painting by Albert Bierstadt, 1850.

8

First Technical Rock Climb

Early Yosemite adventurers like John Muir and George Anderson accomplished many impressive mountaineering feats. However, their equipment and climbing techniques limited the difficulty of the climbs they could attempt. If the technical challenges could be overcome, Yosemite's majestic granite walls presented the ultimate playground to the rogues who were discontent with merely walking on two feet.

THE INTRODUCTION OF MODERN CLIMBING TECHNIQUES

Modern rock climbing with its purpose-designed ropes, hardware, and techniques did not arrive in Yosemite until the 1930s. This new era in Yosemite climbing was ushered in by way of Dr. Robert Underhill. Underhill was a Harvard professor who had climbed throughout the Alps. In the early 1930s, the Sierra Club invited him to California to teach them the skills he learned in Europe.

The rock climbing process that Underhill taught went as follows:

First, the lead climber tied a rope around his waist and began to climb. The follower, anchored to the rock below the leader, "belayed" his partner by bending the rope around his (the follower's) waist. When the leader moved up, the belayer fed out more rope, or "slack." Critical to this system was some form of protection that would secure the rope to the wall. That protection came from a unique piece of equipment called a piton. A piton is a forged metal spike that a climber hammers into a crack or seam in the rock. As the leader ascended the rock, he occasionally hammered pitons into the natural cracks that split the granite. The rope was attached to each piton with a carabiner or sling. If the leader fell, the belayer held the rope tight, and the leader would fall only twice the distance from his last piton. Naturally, that's much better than a complete tumble to the ground. Modern climbing equipment has only evolved slightly from these older techniques, and the overall process is still quite similar to these "old" ways.

The early Yosemite rock climbers were well-educated Sierra Club members from Berkeley, California, including Francis Farquhar, Jules Eichorn, and Dick Leonard. They employed Underhill's European techniques to climb several of the pinnacles and spires that are sprinkled throughout the Valley. The Crown Jewels of their pioneering ascents are the Cathedral Spires, located at the southwest end of the Valley. Their ascent of the steep and difficult Higher Cathedral Spire in 1934 was a pivotal moment in Yosemite's climbing history.

Though these early climbers undertook impressive challenges, the big walls—El Capitan, Half Dome, the Sentinel, and the Lost Arrow Spire—were not even considerations for what was possible. One of the reasons these climbers turned their gazes away from the grand climbs was technical ability; it just did not exist at the time. These climbers were primarily

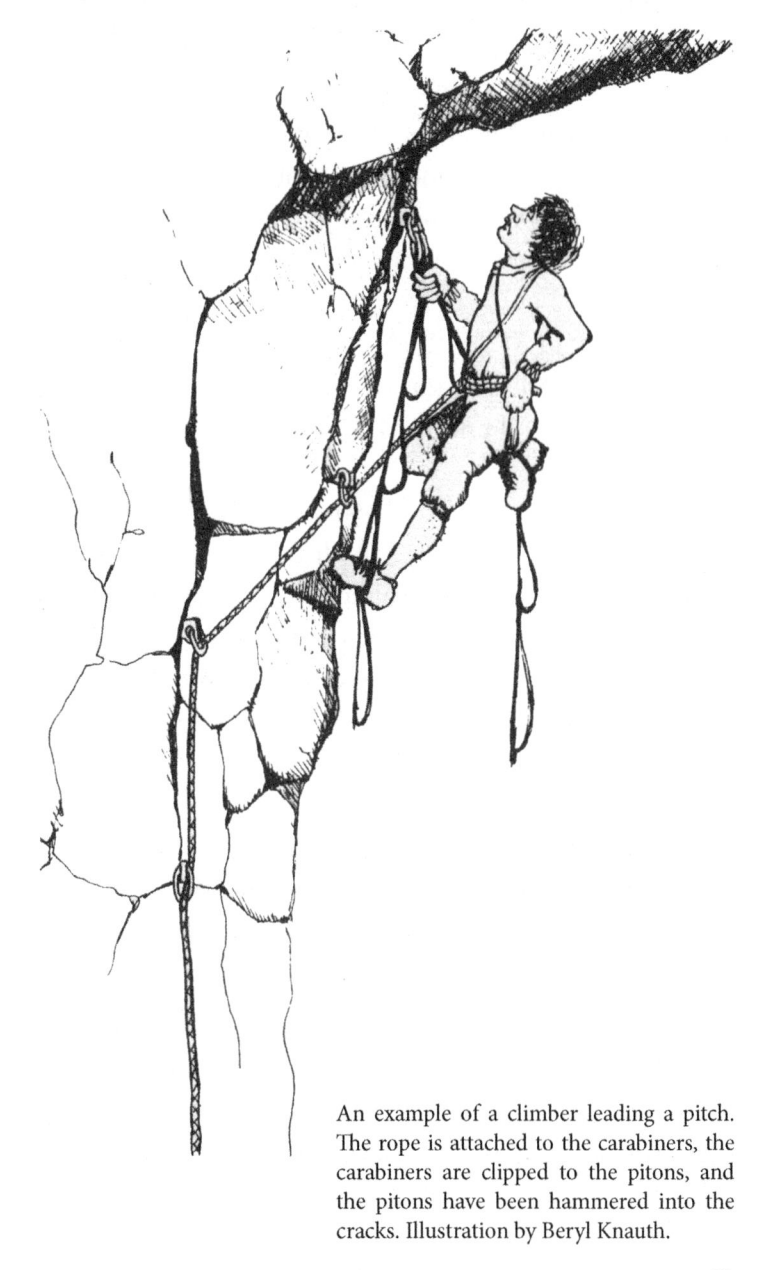

An example of a climber leading a pitch. The rope is attached to the carabiners, the carabiners are clipped to the pitons, and the pitons have been hammered into the cracks. Illustration by Beryl Knauth.

mountaineers and not climbers of sheer granite walls. However, these untouched, long, and committing climbs also required specialized climbing hardware that no one had invented yet.

It would take another decade until the next evolutionary leap in Yosemite climbing took place— at which time the limit of possibility was pushed forward by an unassuming and unexpected character.

LEAFY GREENS, ANGELS, AND HARD STEEL PITONS FORGE THE WAY

John Salathé was born in Basel, Switzerland in 1899. He trained and worked as a blacksmith there until 1929 when he emigrated to San Mateo, California. There, he opened his shop, the Peninsula Wrought Iron Works. He was a hard, nononsense worker. As World War II came to a close in the mid-1940s, health and marriage problems brought on a mid-life crisis for Salathé. He reflected deeply on the life he had lived thus far. During this reflection, he had a spiritual awakening. He decided that to restore his health—both spiritual and physical—he needed to follow a fruit diet, meditate, and spend more time in the outdoors. His explorations into California's wilderness led him to the Sierra Club and rock climbing.

Now in his mid-forties, Salathé was old for a beginner rock climber of that time. Most of his contemporaries were still in college. He more than made up for this difference with his extreme fitness and newfound spiritual conviction.

In later years, he would often speak of the angels who guided his climbing ascents.

When Salathé first visited the Valley as a climber, the only pitons available at the time were from Europe. These European pitons were made of soft iron and were designed primarily for limestone cracks. They worked well in limestone because when hammered they would bend to the shape of the irregular cracks and hold tight. However, these soft pitons did not work well in the granite cracks of Yosemite. Salathé immediately saw this on his early ascents in Yosemite. The pitons often bent over before they could be hammered deep into the cracks. A bent piton was both insecure and ruined for future use. The pitons that climbers hammered deep into cracks often bent in the process, and if that did not destroy their shape, then trying to remove them did. So, for a climber looking to do a long climb, perhaps hundreds of pitons would have to be carried. This was an expensive and laborious prospect which prevented anyone from climbing long, technical routes.

Salathé was a problem solver. One day while climbing, he saw a blade of grass protruding from a crack. "What if," he thought, "there was a piton like that blade of grass?" A piton that could be hammered into the crack, hold its shape, hold firm to protect the climber, and be removed and reused.

To make such a piton, Salathé knew he had to use hard steel. He took an old Ford Model A axle, made from a steel alloy, melted it down, and poured the recycled steel into his piton molds. Now, all he had to do was test them.

THE ULTIMATE TEST

First, he needed a partner willing to undertake such an endeavor. One of his fellow Sierra Club members was ready. Anton "Ax" Nelson was a giant compared to the lean and wiry Salathé. The 27-year old worked as a carpenter in the San Francisco Bay Area. He was strong and determined—the perfect partner.

In November 1946, to test the utility of Salathé's pitons, they turned their eyes to the Southwest Face of Half Dome. This face is the steep shoulder of Half Dome visible from Glacier Point and on the opposite side of the Dome from McAllister's Cables Route.

Climbers had not put up a new route on Half Dome in the seventy-one years since George Anderson climbed up the Northeast Shoulder in 1875, but it wasn't for lack of trying. Dick Leonard, who completed the first ascent of Higher Cathedral Spire in 1934, attempted to climb Half Dome's Southwest Face twice. There is no doubt that his failure was due in part to the inferior soft iron pitons that he used during those attempts, which occurred before World War II and Salathé's arrival in Yosemite.

Salathé and Nelson made the long hike up to the slightly less than vertical Southwest Face. They would have followed parts of Anderson's old trail that skirted Vernal and Nevada Falls. At the mouth of Little Yosemite Valley, they would have hiked over an ancient glacial moraine with unimpeded views of Half Dome's massive South Face. From here they walked behind Liberty Cap and Mount Broderick, around Lost Lake, another remnant of the glaciers, and on towards the "Diving Board" that sits below Half Dome's west shoulder (not to be confused with the Visor, which juts out from the summit).

Many adventurous spirits had already ventured up here before these climbers. Ansel Adams, the most acclaimed photographer of Yosemite's landscapes, came up to the Diving Board in the 1920s. It was here that he captured one of his most famous photographs, "Monolith, the Face of Half Dome."

Perhaps Salathé and Nelson wandered over to the Diving Board and peered over its edge. They may have fantasized about climbing the sheer, 2,000-foot Northwest Face. They wouldn't have lingered long though, for their goal that day was more fathomable for the era and reasonably within their grasp.

At the base of the Southwest Face, they would have unpacked their rope, pitons, and carabiners. They brought about 30 of Salathé's pitons; this was a fraction of the pitons they would have needed if they had used the older style made from soft iron. They had to organize the pitons according to size on a sling. Then the sling went around the leader's shoulder. The leader would have tied one end of the rope around his waist. The follower would have bent the rope 360-degrees around his butt, ready to feed out slack for upward progress or grip the rope tight in case of a fall, thus putting the leader on belay.

Perhaps Salathé and Nelson were anxious. They had only been rock climbing for about a year. They had both accomplished challenging climbs, but what they were about to attempt was far bolder than those previous ascents.

They started up the wall. The climbing was tediously slow. At times, it was easy enough that the leader could place his hands and feet inside of the various cracks to pull himself up. Sometimes, the natural granite holds were not big enough to climb on. In this case, the leader had to hammer a piton into a thin crack no wider than a centimeter. Then he would pull himself upward using the piton and repeat the process. At the end of the "pitch," or the allotted length of rope in use, the leader anchored the rope for the follower to ascend. While ascending, the follower hammered out the pitons to reuse on the next pitch.

To their delight, the new pitons worked! The recycled steel alloy was hard enough to hold its shape after repeated abuse, yet still retained enough "spring" to stay tight in the cracks and give the leader the confidence to quest upward.

The men moved steadily up the face, but night drew near. They realized that they would have to sleep in this vertical world. No one before them had ever slept on a route this technical in Yosemite. Not only were they forging their way up the first truly technical rock climbing route on Half Dome, but they also were making the first bivouac on a Yosemite wall. And Salathé's new pitons made it all possible.

That night, on the ledge, they watched the moon rise over the High Sierra to the east. As they braced against the chill of the autumn night, they would have discussed the pitches they had completed and what might lie ahead in the morning. The climbing had not been particularly difficult. They only needed the right equipment—the hardened pitons. Even though they still had climbing ahead of them the next day, they knew that they had broken through another ceiling of impossibility. They would have wondered, with excitement, what other Yosemite walls they might climb with the aid of Salathé's steel?

The next day, the two men reached Half Dome's summit. They had done it. They had completed the first modern, technical rock climbing route on Half Dome. They had climbed 900 feet of technical terrain. They used Salathé's pitons 150 times, and they didn't drill a single bolt.

Using these same techniques, Salathé went on to make the first complete ascent of the Lost Arrow Spire with Nelson in 1947 and the first ascent of the Sentinel with Allen Steck in 1950. Perhaps unbeknownst to Salathé, he had ushered in the Golden Age of Yosemite rock climbing.

First with Anderson, then with Salathé, Half Dome slowly became the place where climbers came to shatter the impossible. The next significant ascent of Half Dome would be far more daring and break down the psychological barriers to Yosemite's biggest walls; it would also set in place a legendary yet friendly rivalry that many climbers talk about to this day.

Woodblock print by William Seltzer Rice, 1920.

9

THE REGULAR NORTHWEST FACE

By the 1950s, only two of Half Dome's four sides had been climbed: the Northeast Face where the cables had been placed over Anderson's route and the Southwest Face via Salathé and Nelson's 1946 modern rock climbing route. The two that remained were the imposing Northwest Face and the seemingly featureless South Face.

AN IMPOSING WALL OF ROCK

The Northwest Face is Half Dome's most defining feature. It is the "cleft" face—the "missing" half. Tis-sa-ack's tears stain this sheer, 2,000-foot tall granite wall that towers over Mirror Lake and Yosemite Valley.

Climbers had yet to seriously consider—let alone attempt—the Northwest Face. The face is slightly less than vertical at 85 degrees, and the prospect of climbing into this unknown terrain was frightening. Keep in mind, this story takes place in the days before speedy helicopter rescues. In fact, no one had ever performed a rescue on the side of a Yosemite big wall before. Anyone venturing onto the face would do so at their own peril.

Around 1945, climbers started to study the face in an attempt to find a route. They peered at it with binoculars from the Valley floor and hiked to the summit to examine it from above. The steepness and size quickly discouraged them.

In 1954, three climbers, Dick Long, Jim Wilson, and George Mandatory, made the first exploration up the face. They retreated to the ground after only 175 feet of climbing. However, they had discovered the gateway into this expanse of exfoliation flakes.

During the climbing season of 1955, four other climbers were ready to give the face a go. These were Royal Robbins, Warren Harding, Jerry Gallwas, and Don Wilson. Together, they were a team of talented and capable Yosemite climbers.

Royal Robbins was 19 years old. His dark hair was buzz cut, and he wore stylish wayfarer prescription eyeglasses. He was lean and tall, with a face chiseled like that of a Greek statue—in other words, he was quite handsome. Robbins was also a key proponent of the emerging climbing ethic that is still in use today. This ethic stated that routes should be climbed using as few bolts and pitons as necessary because these tools damaged the rock. Climbing should be as "clean" as possible—that is, using the existing features of the rock for upward progress and altering the rock as little as possible in the process.

Warren Harding was 31 at the time, and some of his more youthful peers in the Yosemite climbing scene considered him too old for this sport. Of course, he was a youth compared to John Salathé. Harding was a short man with a nest of thick, wild hair. He made a good wage as a surveyor, and his supervisor allowed him plenty of vacation time to pursue rock climbing. Harding also had a love of sports cars and cheap red wine. He was a stark contrast to Robbins and his ethic of clean climbing. Harding's ethic, if it can be called that, was to climb for himself and have a good, often intoxicated, and laughter-filled time in the process. And when he found a line up a wall that he liked, he climbed it, regardless of how many bolts he had to drill into the virgin stone.

There could not have been a more opposite pairing of climbers than these two hard-headed men. Could they overcome their differences to succeed where few others had even tried?

When the team finally stood at the base of Half Dome, the sight above them would have overwhelmed their senses. From Mirror Lake, there are 3,000 vertical feet of steep slabs before one reaches the bottom of the Northwest Face. To stand at the base of the face, a climber sees 2,000 feet of near-vertical granite above, including the overhanging summit Visor, and 3,000 feet of Valley views below. This combination of immediate exposure below and foreboding wall above will weaken the resolve of the strongest of people. Though fear gripped their minds, the men were excited and determined to get on the wall.

Over three days, they ascended 500 feet of the face. The rock climbing through this section was not particularly difficult, but these experienced climbers moved far slower than their honed skills would have allowed. Doubt and fear tore their will apart. At 500 feet above the base, Robbins and Harding were determined to keep climbing. Harding said:

> "Even if we can't make the summit, let's push the route as high as we can. We have plenty of food and water for another couple of days."

Wilson, the leader of the group, did not share their enthusiasm. The team retreated to the ground. Of the attempt, Robbins said:

> "We crept away from there like whipped curs, with our tails between our legs. We dared what no one else dared, and we were found wanting. I didn't like the feeling, and vowed to return."

THE BIRTH OF A CLIMBING RIVALRY

Robbins wouldn't return to the face for another two years. During this time, he and Gallwas planned and prepared. They refused to accept failure a second time, so their planning had to be meticulous. Gallwas forged new pitons for the ascent. He used Salathé's pitons as models. Like Salathé, he forged his from chrome-molybdenum steel that resulted in a hard, reusable piton. In their planning, Robbins and Gallwas knew they would encounter cracks too wide for the largest pitons of the day. So Gallwas made new, bigger pitons that could fit inside a two and a half inch wide crack.

Robbins and Gallwas recruited a third partner who was new to Half Dome, Mike Sherrick.

But what about Warren Harding? Harding, too, had not forgotten about Half Dome. He had also recruited a team of talented climbers: Mark Powell and Bill "Dolt" Feurer.

Why did Robbins, Gallwas, Wilson, and Harding not join each other for the second attempt? Like all climbing partnerships that dissolve, personality differences are often the heart of the matter. Robbins and Harding were just too different for a good pairing, both in their climbing ethics and their approaches to life. Robbins was proper and disciplined. Harding was farcical and irreverent.

Harding had been working in Alaska and returned to California in the summer of 1957. He immediately wrote to Powell and suggested that they attempt Half Dome soon. However, Robbins had already heard that Harding would be back in California and was hungry for Half Dome. He called Gallwas and told him that the time for Half Dome was now. Gallwas later recalled:

> "The following week while I was studying for finals, the phone rang and quite to my surprise, it was Royal. His message was simple. Warren was returning from Alaska in a few weeks and intended to make an attempt on Half Dome with Mark. While we had not been planning a serious attempt at that time, it seemed clear that Warren and Mark were. We concluded that we had better plan and act swiftly."

AN ASCENT FOR THE HISTORY BOOKS

Robbins' team blasted up the Northwest Face on June 24, 1957. A couple of days later, Harding and his team arrived in the Valley ready to attempt the face. To their amazement, they saw that a team was already up there. Harding knew Robbins, and he knew that failure for Robbins was unlikely. His heart must have sunk in that moment. A historic ascent of Half Dome had been within his grasp, and then it was snatched away. Harding was not a man to ruminate long on his failures. Instead, his desire quickly moved to the grandest of Yosemite's walls—a wall so grand that no one else had even entertained the idea.

Meanwhile, Robbins, Gallwas, and Sherrick had hauled five days of food and water up the face. By the end of the first day, they reached the high point from the previous attempt.

They slept on a narrow ledge, still wearing their makeshift harnesses. Tethers went from their harnesses to a few pitons hammered into the natural cracks; this was all that held them to the wall.

The easy climbing was now below them. What lay ahead would be far more difficult. There were featureless sections of rock, disconnected ledges and cracks, and the intimidating Visor that loomed overhead and blocked their way to the summit.

At one of the blank sections, rather than drill more bolts into the rock, and thus scar it for eternity, Robbins attached his rope to a lone piton. Sherrick, his belayer, lowered Robbins 50 feet down the wall. Then Robbins began to run back and forth across the wall. Robbins' goal was to swing himself over to a natural rock chimney where they could continue their upward progress. There was nothing but nearly a thousand feet of air below him. Running across the wall was a nerve-wracking

Royal Robbins leading a pitch on the Northwest Face of Half Dome during the first ascent. Photo from the Jerry Gallwas Collection and the Yosemite Climbing Association.

effort for Robbins and his companions. After several attempts, Robbins reached the chimney. They could continue! But the sun was quickly setting over El Capitan to the west. They set up camp on a nearby ledge and settled in for another night in this vertical wilderness. Their sleep that night must have been fitful at best for great uncertainty loomed overhead.

The following three days brought challenging, and often scary, pitches. The Northwest Face of Half Dome is known for the poor quality of its rock. Robbins' team was experiencing this fact firsthand. The face of Half Dome was—and still is—literally falling off, one giant flake at a time. This is because of the geological process of exfoliation, as explained earlier. Robbins, Gallwas, and Sherrick weaved a route toward the summit that followed these loose flakes and exfoliation cracks. Why would they choose such a route? Because it was the easiest and most intuitive path.

Two of the rock features they climbed through have since fallen off of the face. The Robbins Traverse, where Royal swung like a pendulum into the chimney, broke free sometime in 2015. Higher up, Psych Flake fell from the wall a decade after Robbins' team carefully climbed behind this towering knife blade of rock that moved when the climbers pressed on it.

Rock climbing routes are not static, unchanging vertical paths. The Northwest Face of Half Dome is evidence of this, and the three men took care as they moved up this vast sea of loose rocks with nothing but thin, fragile ropes holding them to the wall.

Though they were now in their third day on the wall, the climbers were not without their doubts. Robbins remembered:

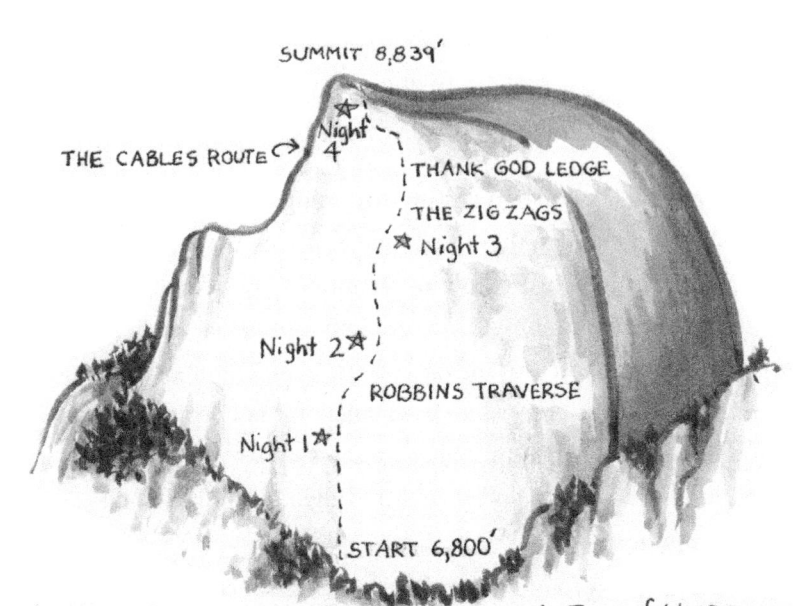

Robbins' Route up the Regular Northwest Face of Half Dome

by Susan Greenleaf

> "We feared the enormity of the wall ... We dreaded having to reach so deeply within ourselves and maybe find ourselves lacking."

Perhaps conscious of the danger that surrounded them at all times, Robbins said of a rescue, should it be required:

> "It would take days for rescuers to reach us ... A rescue like that had never been attempted, for the very good reason that a wall like Half Dome had never been climbed."

Despite their uncertain destiny, they moved higher up the wall. Now over halfway, they had found a rhythm to the climbing. The team used two climbing techniques to ascend: free climbing and aid climbing. In free climbing, the leader only uses his hands and feet for upward progress, and pitons, while hammered into the cracks, are used solely for fall protection. In aid climbing, the natural holds are too small, so the leader beats pitons into cracks and seams, and then pulls himself up using the pitons.

The fourth day found them at the Zig-Zags: a series of thin, two hundred foot tall cracks and corners that Gallwas had to aid climb. Above them, the overhanging Visor both beckoned and intimidated. And just below the Visor, they could see another blank section of rock. They still had no clue as to how they would overcome this final obstacle guarding the summit.

When Gallwas reached the top of the strenuous Zig-Zags, he saw their salvation—a narrow ledge traversed underneath the blank headwall. They scooted along this ledge, dubbed

Thank God Ledge, and then bivouacked for the night.

The next day was the 28th of June, 1957. The team had been on the face for five days. They climbed several more pitches, and near sunset climbed onto the summit of Half Dome. They called the route the Regular Northwest Face. It was the longest, steepest, and most technical rock climb ever accomplished by American climbers up to that date.[3]

Who was there to congratulate them with sandwiches and wine? Warren Harding, Mark Powell, and Bill Feurer, the climbers who had barely missed being the first up this face. Harding was happy for his friends and fellow climbers in their accomplishment. However, he was not a person who accepted defeat readily. Harding would come back to Half Dome one day and in search of a new route, but it would not be without its trials. In the meantime, feeling robbed of Half Dome, he turned his attention to the only object worthy of his vision: El Capitan.

Robbins, Gallwas, and Sherrick's ascent of Half Dome meant that climbers had scaled three of Half Dome's four sides. All that remained was the almost featureless South Face that soars above Little Yosemite Valley.

The view from Mount Starr King. The Southwest Shoulder and South Face of Half Dome are on the right. Liberty Cap is below Half Dome. Upper Yosemite Falls is visible left of center.

10

THE SOUTH FACE

After losing out to Royal Robbins, Warren Harding turned his full attention to the only other unclaimed trophy in Yosemite Valley: El Capitan. By 1957, a few people had probed up this sea of granite, but no one had committed to the 3,000-foot face that included vertical and overhanging rock. While many climbers wanted to ascend this monolith, the challenge was too big and intimidating.

Harding started up El Capitan within days of Robbins reaching the summit of Half Dome. However, Harding's ascent would be nowhere as fast. His team climbed the Nose of El Capitan in siege style over the course of 18 months. They finally reached the top in the fall of 1958 after spending a combined 47 days on the wall.

His ascent of El Capitan was a considerable accomplishment. However, it did not replace Harding's yearning for Half Dome. After Robbins' successful ascent, Harding said, "My congratulations were hearty and sincere, but inside, the ambitious dreamer in me was troubled."

ATTEMPT TO RECLAIM A LOST SUMMIT

While Harding made haste in marking his claim to El Capitan, his return to Half Dome would take nearly a decade.

By the summer of 1966, Harding, now over 40 years old, and his friend Galen Rowell, a mechanic turned National Geographic photographer, had found a new route to attempt on the Dome. This route went straight up the middle of the South Face.

The South Face of Half Dome rises 2,200 feet above Little Yosemite Valley. Two massive waterfalls and another 2,000 feet separate the peaceful Little Yosemite Valley from the hustle of Yosemite Valley below. The Merced River quietly flows through this broad, glacier-carved valley before cascading over Nevada and then Vernal Falls.

The start of the South Face route follows a massive, 1,000-foot tall rock arch. This is the Great Arch, and it gracefully sweeps into the middle of the face. It is the distinguishing feature on a face that otherwise appears blank and unclimbable. Above the arch, there are a few stained water grooves and large solution pockets—places where water has eroded softer rock to create a natural pothole.

While this section of rock was quite aesthetic, there was one problem. There were only a few natural cracks above the Great Arch, and they were discontinuous. Other climbers, too, had looked at the line. Unlike Harding, they dismissed it because it would require too many unnatural bolts to be drilled into the rock to gain upward progress. A firm ethic had taken hold of American rock climbing: alter the rock as little as possible. A bolt was the ultimate scar. Harding and Rowell estimated that 25 percent of the route would require bolts. For most climbers, this was unacceptable, but not for Harding. He didn't care; he climbed for himself. He had found a beautiful piece of rock and wanted to climb it.

The South Face of Half Dome. Note the Great Arch on the lower right. Photo by Walter Siegmund.

Harding and Rowell recruited Yvon Chouinard and Chuck Pratt, two of Yosemite's best climbers, to join their first attempt in the sweltering summer heat. They started up the Great Arch, but the first pitch of climbing presented a setback that would be foreboding for this climb. Chouinard dislocated his shoulder, a painful and slow healing injury. With this, Chouinard and Pratt decided to abandon their role in the climb.

Rowell also wanted to abort the mission for the time being, but Harding insisted that they continue up. Over the course of three days, they climbed 900 feet up the arch. Near the top of this feature, they found an overhang beneath which they hung their hammocks. Just as they were settling into their vertical campsite, a massive summer thunderstorm moved in. Rain fell throughout the night, and lightning illuminated the valley below. They hoped the storm would abate before long. After 36 hours of hanging in their hammocks, the rain continued to fall. Finally, they accepted defeat and rappelled to the ground.

The two friends wouldn't get back to the route on Half Dome for another two years. Harding took a construction job in Vietnam; with the war there in full swing, it turned out to be a more dangerous job than he had anticipated.

FLIRTING WITH DISASTER

By November of 1968, they were ready to head back up the South Face. It was an unusually warm and dry fall in the Sierra Nevada, and they thought the conditions perfect for another attempt up the unclimbed face.

They reached their previous high point of 900 feet up the

arch after three days. That night a storm moved in, but by the morning the skies were clear and the rock dry. They continued up.

The fourth day they exited the Great Arch and started up the mostly featureless slabs of the upper wall. Their progress on the upper wall was slow. Without natural features to climb, they had to drill holes and place expansion bolts in them. Occasionally, Harding would drill a shallow quarter-inch hole, set a hook-shaped piece of iron on the bottom edge, and use that for progress. This too was slow, but faster than drilling one to 2 inches into the rock for a bolt.

The end of the 6th day brought them about 1,500 feet up the face. They talked via walkie-talkie to a friend on the ground to get a weather report. The forecast called for cloudy skies, but not storms. "Perfect," they likely thought. The two climbers hoped to be on the summit in another three days. They settled into their constricting hammocks, ate dinner, and fell asleep.

The weather in California's Sierra Nevada tends to be agreeable—i.e., sunny and predictable. These are excellent conditions for rock climbing, but they breed complacency among climbers. Many assume that each day will be sunny, and if there are clouds, then nothing will come out of them. But big storms do sweep through the Sierra, often settling in for two or three days. These storms can drop lots of moisture in any of its forms in any season. In Yosemite, rain wicks off the exposed granite slabs and shallow soils. Ephemeral waterfalls form instantly—jetting off cliffs and rushing down slabs.

At midnight, rain started to fall on Harding and Rowell.

Warren Harding surviving in his hammock on the South Face of Half Dome. Photo by Galen Rowell.

Small drops landed on their shoulders and heads, but those drops increased to a steady rainfall as the morning drew nearer. Their nylon hammocks filled with water. Their clothes became soaked.

That morning, when they peeked out from their hammocks, a layer of snow covered everything in sight: Little Yosemite Valley, Mount Starr King, Mount Clark, and peaks of the High Sierra in the distance. The ponderosa pines below were frosted with snow, and even the wall had a crusty layer of snow plastered to it.

The snow above them started to melt in the warmth of the day. A fresh dousing of meltwater cascaded down the rock. Even with this second soaking, they weren't too concerned. They knew that later in the day the sun would come out. It would dry their gear and the rock, and they could continue with the climb.

Unfortunately, the weather conditions deteriorated through the rest of the day. The snow kept falling and growing in intensity. Small avalanches sloughed washing machine-sized chunks of snow and ice off the face above them. The sun never broke through the clouds, leaving Harding and Rowell still wet and cold as the second night of the storm set upon them.

In November, the days are short and the nights are long. They passed the second night without sleep—14 hours in the darkness—shivering and exhausted, fighting off hypothermia. Their fingers and toes were numb; their skin purple and raw.

With dark bags under his eyes and his hair exceptionally disheveled, Harding looked out from his hanging tent on the morning of the eighth day to find the landscape transformed into even more of a winter wonderland. The temperature had dropped, and snow continued to fall. Ice clung to the rock. The small avalanches sliding down the face had grown in intensity and were battering them in their exposed position on Half Dome.

They shivered uncontrollably. For Rowell, his hammock had become a straight-jacket that would soon be a coffin if he didn't do something. Images of his wife, children, and home flooded his mind. It was Sunday morning—what he wouldn't give to be at home on a quiet Sunday morning standing by the heater with a cup of tea and a hot plate of bacon and eggs. Reality set in. They had to descend.

Rowell attempted to rappel down the ropes. The rain and then freezing temperatures had left the ropes frozen and caked in ice. A veneer of glassy ice covered the rock face. He soon realized that descending 1,500 feet in these conditions would be impossible. He abandoned the effort and ascended back to their anchor where Harding waited.

Harding helped Rowell set up his hammock and get back inside. They took stock of their situation; it was a grim predicament. They were hypothermic and entirely exhausted—physically and mentally. How much longer could they last before succumbing to the elements?

They heard shouts from below. Harding turned on the walkie-talkie. It barely worked, and the sound was scratchy, but they heard their friend on the other end.

Harding said into the radio, "We're really not doing too well … it's cold and wet and we're a little numb … get us off, somehow." Harding had just asked for a rescue, the first and last of his life. This was a hard defeat to accept, but they wouldn't make it through the night without help.

That afternoon, as the sun approached the horizon, they heard the tell-tale *thawp thawp thawp* of a helicopter's rotors. The helicopter came up from Yosemite Valley, over the Porcelain Wall, and between Liberty Cap and the Southwest Face of Half Dome. It drifted toward them, appearing as a dragonfly hovering closer and closer, but just out of reach. After one fly by, it disappeared over the summit of Half Dome.

The helicopter flew by again a few minutes later. A spool of rope swung below it. They thought the helicopter would drop the rope to them, but it flew out of sight. This reappearing and disappearing act played out several more times before sunset. Then a cold and hopeless darkness washed over them.

They knew that the helicopter wouldn't fly through the mountains at night; it was too dangerous. They wondered what was happening with the rescue and why a rope hadn't been dropped to them yet. Finally, Harding said what they were both thinking, "We're going to spend another night up here."

It was their third night trapped at this lonely anchor on the massive face. Harding knew that neither of them would last two or three more hours. They dutifully climbed back into their hammocks as the mercury dropped. Their wet clothes started to freeze. They had accepted that this place could be their tomb, but the moon was out that night. They had an unparalleled view of a beautiful world above and below them. The Clark Range faded over the horizon to the south. The

Cascade Cliffs dropped into the Merced River Canyon. The sound of Nevada Fall echoed from the foot of Liberty Cap.

While Harding and Rowell worried about the competence of their rescuers and the helicopter pilot, the first large-scale, technical big wall rescue in America was being organized. The helicopter ferried nine rescuers, ropes, tents, dry clothes, and warm food to the summit of Half Dome.

On Yosemite big walls, there is a limited range of sounds that a climber will hear. The grunts and bodily noises of their partner. The snap of carabiners and tap of the hammer. The swoosh of swifts—fast, slender, cliff-dwelling birds—dive-bombing down the wall. A roaring waterfall. And relentless wind. Sometime after dark Harding and Rowell heard a new and strange sound. They looked out from their hammocks to see someone rappelling down a rope.

"Are you one of the guys from that chopper?"

The man was wearing a down parka and carried a walkie-talkie and a large pack. A head-lamp lit up his descent. He was self-assured and moved with swift confidence. The rescuer's identity revealed itself in Rowell's mind. Harding was still putting it together.

Harding leaned into the rescuer's hooded face. "Who are you, anyway?"

It was Royal Robbins.

Royal brought dry down parkas, warm gloves, and hot soup. The two climbers put on the dry clothes. They relished the warm soup which brought needed energy and comfort to their bodies.

They then ascended Royal's rope to the summit where more rescuers and food waited. That night they camped on top of Half Dome. They sipped on brandy, ate firefighter rations, and slept comfortably in tents. The helicopter returned in the morning. Harding and Rowell caught the first flight down to the Valley, catching sight of their abandoned equipment dangling in a frozen vertical sea of ice, snow, and rock.

For some, this defeat would be the end of the affair. A smarter mountaineer might abandon the climb for good, but Harding and Rowell were transfixed.

TRAGEDY AND A HAUNTING CLOUD OF FAILURE

They scheduled the next attempt for September 1969, but misfortune beckoned. Harding was struck by a car while crossing a street the night before they were to start their attempt. Rowell found Harding in the hospital with a demolished leg. The doctors told Harding that it would be at least a year before he would climb again, if ever. They even questioned if he would ever walk again.

In December of 1969, Harding had surgery to reconnect cut ligaments and insert pins to hold the bones in his leg together. Many of his friends thought that his situation was grim, but Harding was unperturbed. He wanted to get back on the wall.

They attempted the route two more times in June of 1970, but storms beat them back.

Harding climbing through the steep roof of the Great Arch. Photo by Galen Rowell.

Harding's route up the South Face of Half Dome

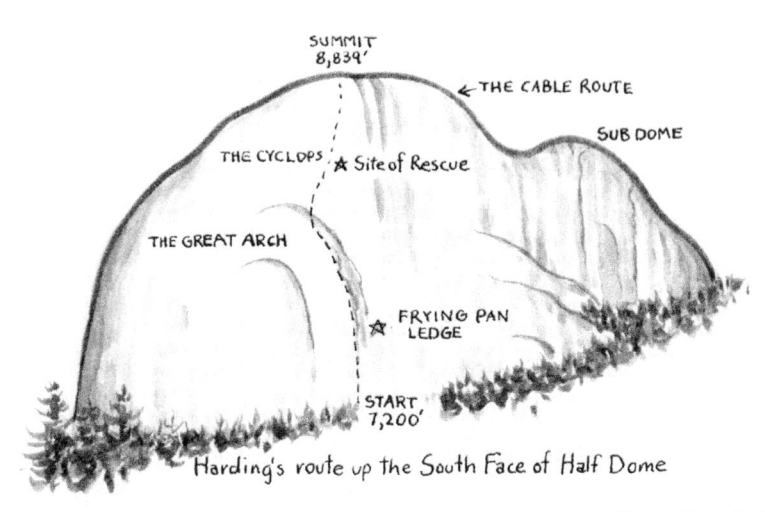

by Susan Greenleaf

REDEMPTION BY PITON

On July 4, 1970, they started up the route once again. It was their sixth attempt, and they were determined never to climb the route again, regardless of the outcome. That night there was a new moon. The only light to touch the face came from the dim twinkling of stars, and none of that reached under the Great Arch where Harding and Rowell bivouacked. After so many failures, what went through their minds? Perhaps it was the abandonment of dichotomies such as success vs. defeat and conqueror vs. conquered.

Harding had already come to terms with abandoning these staid alpine concepts after climbing the Nose on El Capitan. Now it was Rowell's turn. In his account for the 1971 *American Alpine Journal*, Rowell wrote:

> "Here on Half Dome our five defeats versus one still questionable success can hardly be rationalized into a 'conquering.' What we have shown is an extension of man's greatest natural gift: his adaptability. The ecology movement is beginning to remind man that in order for a species to survive it must adapt not only its physical characteristics but also its behavior to its surroundings. Man has turned the tables. He is trying to adapt his surroundings to himself. Climbing is an activity in which man works in surroundings far less adaptable than normal. Is the climber trying to adapt them to his needs … to conquer them? I think

not. Possibly he finds something unconsciously satisfying in returning to a biologically proven situation where it is he who becomes adapted."

After three days, they passed their previous high point and the gear they had abandoned, a reminder of their close encounter with death. The sky hinted at a storm beyond the horizon. Moisture-rich cumulus clouds followed the high altitude cirrus clouds that herald a warm front. The high country to the east was dark and gray. Rain fell over Tuolumne, Cathedral Peak, Matthes Crest, and Mount Lyell. Nightmares of another retreat swirled in their minds, but the two climbers continued up.

The tedious climbing on the upper slabs continued over the next three days, as did the threat of a storm.

On the sixth day, fat, fluffy cumulus clouds moved around them, hiding and then revealing Half Dome's summit to visitors in Yosemite Valley below. A light rain fell on the climbers. Harding climbed in a down parka and a cagoule—a long, waterproof jacket.

Finally, on the afternoon of July 9, 1970, the *tap tap tap* of the hammer hitting a drill or a piton—a sound that floated through the air of the South Face for so many days—stopped for good. Harding yelled, "I'm up!" Their five-year obsession was over.

The hard-won climb had led Rowell to reflect deeply on the meaning of what they had done. Galen Rowell was a part of the modern conservation movement. He saw the pressures that the increasing popularity of the outdoors was putting on the environment. At the same time, he saw the demands that a ballooning world population placed on individual people, and how some of those people now sought solace in the outdoors.

He was not alone in these thoughts. Yosemite Valley was no longer a quaint retreat. Half Dome was no longer an impossible summit; it was attainable by almost any hiker. The crowds that flooded into Yosemite overwhelmed the infrastructure and the environment, including Half Dome and its surrounding forests. Humanity had to do something if this landscape was to remain pristine for future generations.

11

Managing a Mountain

The need to control our landscape and manage it seems to be a natural inclination for humans. To understand how the National Park Service manages Half Dome, we must go back in time more than a century, to the creation of Yosemite National Park.

A MISMANAGED STATE PARK

Abraham Lincoln signed the Yosemite Grant in 1864, establishing Yosemite Valley as a public park for all generations. Although the federal government set this land aside for public enjoyment, the land was handed over to California to manage as a state park.

John Muir, a stalwart defender of the Yosemite landscape, continued to lobby the government for protection not only of the Valley but also of the surrounding area, including Tuolumne Meadows and the high country of the Sierra Nevada. His efforts to protect this vast swath of wilderness succeeded in 1890 when Congress created Yosemite National Park, which covered the upper drainages of the Tuolumne and Merced Rivers. As a National Park, these lands fell under the management of the federal government, and the responsibility

Shuttle buses at Glacier Point. Photo courtesy of the Yosemite National Park Archives.

The proud owner of a Lincoln Zephyr stands in front of the Ahwahnee Meadow and Half Dome. Photo courtesy of the Yosemite National Park Archives.

of administering these new areas went to the United States Cavalry. However, California still managed Yosemite Valley.

By the late 1890s, the Cavalry was succeeding in its mission to protect and preserve the new park. California, however, was failing to provide proper stewardship for Yosemite Valley and the Mariposa Grove of giant sequoias. The state-appointed commission that ran the Park was accused of corruption and mismanagement. The commission was regarded as seving the interests of the concession companies, which made handsome profits through hotels, snack shops, tollroads, and toll-trails. Many people saw this as exploiting public land for private gain. Additionally, the giant sequoias were at a constant risk of being logged.

Again, John Muir came to the rescue. He petitioned President Theodore Roosevelt to reclaim control of the Valley and the Mariposa Grove, and thus achieve unity in Yosemite National Park's management. Roosevelt was convinced, and in 1906 he brought the Valley under the control of the National Park.

Curiously, while Congress was steadily designating National Parks, a dedicated branch of government to manage these parks did not exist at the time. A group of committed and influential conservationists recognized that the nation's parks needed to have independent management from other public lands where resource exploitation directed management policies. They slowly put together their plan and began to lobby Congress. The work of industrialist Stephen Mather, lawyer Horace Albright, the Sierra Club, and the General Federation of Women's Clubs paid off in 1916 when President Woodrow Wilson signed the National Park Service Organic Act.

THE BIRTH OF THE NATIONAL PARK SERVICE

The Organic Act of 1916 created the National Park Service and gave it the following mission:

> "The service thus established shall promote and regulate the use of the Federal areas known as national parks, monuments, and reservations ... to conserve the scenery and the natural and historic objects and the wild life therein and to provide for the enjoyment of the same in such manner and by such means as will leave them unimpaired for the enjoyment of future generations."

The Organic Act was particularly eloquent for a government document, and it established a high level of protection for the parks. However, in its eloquence, there was considerable ambiguity. For example, could Yosemite build more roads to facilitate public enjoyment of the land? Could the Park allow the concession company that operates the hotels and restaurants to construct an aerial tram from the Valley to Glacier Point because it would provide excellent views and easy access for visitors? Could the Park allow the same concession company to build a hotel atop Half Dome, just as George Anderson dreamed in 1875? Arguments were easily made that these "improvements" would benefit current and future generations of park users. However, they were not in keeping with the conservationist ethos that the land should remain undeveloped—the mark of humanity being a simple trail through the forests and mountains.

Eventually, various officials and businessmen proposed all of these ideas. The modern conservation movement that took hold in the 1960s recognized that the Organic Act was not enough to protect America's public lands. There needed to be another layer of protection to keep park management from inflicting lasting harm on the landscape by bowing to the will of private companies and powerful, but small interests who wanted to see the wildest spaces paved over with asphalt.

THE WILDERNESS ACT—
PROTECTING THE PARK FROM THE PARK SERVICE

In 1964, Congress passed the Wilderness Act. This act was notable for several reasons.

- Like the Organic Act before it, the writing style broke from typical congressional statements and presented the new laws in simple and almost poetic prose.
- It defined the idea of wilderness.
- It expanded upon the idea of why wilderness was important.
- It listed specific activities that would not be allowed in designated wilderness.

Here is some of what the Wilderness Act says:

"In order to assure that an increasing population, accompanied by expanding settlement and growing mechanization, does not occupy and modify all areas within the United States and its possessions, leaving no lands designated for preservation and protection in their natural condition, it is hereby declared to be the policy of the Congress to secure for the American people of present and future generations the benefits of an enduring resource of wilderness."

"A wilderness, in contrast with those areas where man and his works dominate the landscape, is hereby recognized as an area where the earth and its community of life are untrammeled by man, where man himself is a visitor who does not remain."

"There shall be no temporary road, no use of motor vehicles, motorized equipment or motorboats, no landing of aircraft, no other form of mechanical transport, and no structure or installation within any such area."

With this law in place, it would take another twenty years for Yosemite's wildest places, including Half Dome, to be declared "Wilderness." That declaration came in 1984 with the passage of the California Wilderness Act, which set aside 89%

of Yosemite as designated wilderness—over 677,000 acres of forests, mountains, streams, and lakes—including everything 200 feet above the floor of Yosemite Valley. There would be no more dreams of hotels perched on the summit of Half Dome. There would be no new lines of cables snaking up the shoulder to the summit, though the existing set could stay because it predated the Wilderness Act.

TOO MUCH LOVE FOR A FRAGILE LANDSCAPE

The designation of most of Yosemite as wilderness—and the new guidelines that came along with it—arrived just in time. Starting in the mid-1960s, Yosemite saw massive increases in the number of backpackers and mountain climbers. From 1968 through 1975, overnight use in Yosemite's remote backcountry increased 250 percent! The peak of backpacking came in 1975, with backcountry visitors in Yosemite spending over 219,000 nights outside. Visitors damaged fragile resources. Wildlife became human-conditioned. Popular wilderness areas such as Little Yosemite Valley, which serves as the base camp for many Half Dome hikers, saw hundreds upon hundreds of visitors per night. Soon, in addition to the throngs of cars on the Valley's roads, there would be human traffic jams on the historic Cables Route. There was nothing wild about the situation.

As the number of people wanting to scale Half Dome—whether overnight or in a day—grew substantially, the National Park Service was forced to confront the challenges of a degraded and congested wilderness experience. The

Wilderness Act required "opportunities for solitude," but these opportunities were increasingly rare in Yosemite, especially on the trail to Half Dome.

The Park Service responded in several ways over the following decades. The first action came in 1972 with the creation of a quota system to limit the number of people camping in Yosemite's wilderness. With this system in place, the Park could begin to fix some of the environmental damage caused by previous overuse. Knowing how many people would be using the area in the future, the Park could also mitigate future impacts by establishing designated camping locations and consolidating trail networks.

By the early 2000s, the Park also had to confront the issue of over-crowding on the historic Cables Route. When Anderson first climbed the Northeast Shoulder in 1875, it was one of the most challenging routes in America. The 1919 cables, however, transformed it into a well-established trail that any person of average fitness could reasonably attain. Of course, in 1919, learning of Yosemite, getting there, and then finding the correct path to the Dome was still a considerable challenge. Over the decades, word of the Cables Route spread around the United States and the world, as did images of the iconic Half Dome. As the mountain became famous around the world and accessibility to Yosemite became easier, more people wanted to climb the route, and justifiably so, for it is a fantastic sight. During the height of the summer seasons of the early 2000s, there were up to 1,200 people a day ascending the cables! Imagine George Anderson's dismay from the grave; if only he had installed a permanent set of cables in 1875 rather than a homemade rope attached to rickety eyebolts, he might have died a wealthy man from tolls collected, and his hotel might have become a reality.

The over-crowding on the cables brought other problems aside from limited solitude. The hordes of people on the summit damaged the delicate ecosystem. Hikers eager to build rock shelters and rock art, e.g. cairns, threatened the existence of the Lyell salamanders that live atop Half Dome. Vegetation was trampled and destroyed. Human waste and toilet paper dotted the summit plateau. Campers destroyed the pine trees that John Muir extolled in his writing, save for one secluded far from sight.

The crowds were also a danger to themselves. The Cables Route had become one of the most dangerous traffic jams in the world. There were deaths on the Cables Route nearly every year. Hikers slipped from the cables and fell to their deaths hundreds of feet below. Often, people became trapped on the cables during storms, which, in the mountains, can seemingly appear out of nowhere. The already polished granite became a slick, glassy surface in these conditions. Worse, the steel wires became giant lightning conductors on the exposed face. With potentially hundreds of people on the Cables Route at once, it became extremely difficult for hikers to descend the cables to safety when the skies opened with rain, hail, or lightning.

The Park responded with a Half Dome quota system in 2010. By limiting the number of people hiking to the summit, they hoped to limit environmental damage, preserve the wilderness ethos, and reduce risks to hikers. Today, the number of people who try to ascend to the summit is limited to 300 per day. This quota is controlled by a permitting system that includes people going to the summit in a day and people going to the summit as part of a backpacking trip. Since the adoption of this permit system, the number of deaths has dropped significantly. During the summers from 2013-2017, there were no deaths on the cables, a statistic that bucked the emerging trend of the prior decade and has resulted in fewer tragedies and grief-stricken families.

While the Park does try to mitigate accidents involving hikers through permits, signs, and education, the fact remains that the mountains are dangerous and people will continue to venture into them.

Half Dome in the winter from Snow Creek.

12

A Dangerous Mountain

Rescues happen on or near Half Dome every year. Some are quite dramatic, like when fellow climbers rescued Harding and Rowell from the South Face. National Park Service rangers regularly rescue climbers from Yosemite's walls, often using a helicopter. However, these high-adrenaline rescues are not the norm. Most rescues simply involve Search & Rescue personnel hiking to the patient (who is more often a hiker than a climber) and helping the injured party back to medical care in the Valley.

Deaths are rare on Half Dome considering the volume of people who hike and climb their way up the monolith. That said, there would be far fewer deaths if people heeded nature's signs, practiced situational awareness, and listened to their bodies. From 1930 to 2018, here are the fatalities:

Rock Climbing:	8
Fall from the Cables Route:	7
Suicide:	6
Fall not on the Cables Route:	3
Lightning Strike:	3
Other or Unknown:	3
B.A.S.E. Jumping:	2

The deaths cataloged here occurred on or near Half Dome. Excluded are deaths that happened on the Mist Trail. Of the thirty-two deaths listed, twenty-eight were males and only four were females. Two of the deceased were in their teens, sixteen were in their twenties, six were in their thirties, five were in their forties, one was in his fifties, and one was in his eighties. The average age is thirty-three, though excluding the eighty-six year-old outlier brings the average age down to thirty-one.

The most likely demographic to die on Half Dome are people in their twenties. Males are far more likely to die than females. Males rock climbing, especially unroped, have the highest odds of dying.

Concerning the danger of the Cables Route, which is by far the most popular way up the Dome, only two deaths have occurred since the Half Dome Permit System went live in 2010. Of the deaths that were a result of a fall from the cables, five occurred during bad weather or when the rock was wet, one was during high crowding before the permit system was implemented, and one was likely health-related. They are not listed here, but several people have survived falling from the cables. They owe their survival to luck, like pants catching on a rock protrusion.

What conclusions can we draw from these fatal numbers? Don't go up the mountain during or immediately after a storm. Don't go up the mountain if you are in poor health. Good preparation is essential for success in the mountains. Preparation includes checking the weather, being physically fit, and having the appropriate gear.

Half Dome after the first snowfall of the season. Courtesy of the Yosemite National Park Archives.

Concerning suicide, and regardless of what one believes about suicide, committing this act in a remote and dangerous location like Half Dome puts the lives of rescuers at unnecessary risk. Rescuers must rappel into hazardous situations. Helicopters, which are prone to crashing, must place rescuers in these dangerous areas. Additionally, bystanders, including rock climbers who are on nearly every Yosemite wall every day of the climbing season, must bear witness to the act and live with the memory for the rest of their lives.

THEIR STORIES

Here are the reports of the thirty-two known fatalities to have occurred on or near Half Dome.[4]

May 2018 | Asish Penugonda, 29
Penugonda and a friend were on the Cables Route during a thunderstorm. Penugonda slipped on the wet granite and fell to his death.

November 2015 | Angela Uys, 26
Uys and two partners set out to climb a popular route on the Southwest Shoulder of Half Dome called Snake Dike, near Salathé and Nelson's 1946 route. They had a late start and were moving slowly because of a party above them. The team decided to retreat by rappelling the route. In the process of preparing to descend, Uys became untethered from both the rope and the anchor. She fell to her death. All deaths in the mountains are sad, but this one is especially so because better knowledge of the equipment could have prevented her death.

September 2011 | Markus Praxmarer, 48

Praxmarer and his partner, Thomas Wanner, were climbing the Northwest Face. They were experienced climbers who had done El Capitan, and they worked as mountain guides in Europe. It was a busy day on the route. There was one party ahead of them and two parties below them. At pitch 7, Praxmarer led the way. The route he chose went through a series of large, loose flakes, a common sight on this exfoliating side of Half Dome. Praxmarer pulled himself onto one of these rock flakes, and it then came loose. The climber fell with the rock. When his rope stretched tight, the rock completely severed the rope. Praxmarer fell 700 feet to the ground. It was a tragic accident, but this is a reminder to climbers that loose rock can never be trusted.

July 2011 | Hayley LaFlamme, 26

The granite of the Cables Route was still wet from a thunderstorm that had rolled through the area earlier in the day. A ranger was stationed at the base of the cables and warned all of the hikers of the dangerous conditions above. Some heeded the warnings, and others continued. LaFlamme and three others in her party reached the summit and then started their descent down the still wet cables. She was about halfway down the route when she lost her footing and fell 600 feet to her death.

August 2011 | Ryan Leeder, 23

Leeder reached the summit via the cables late in the afternoon. Two thousand feet below the Visor that hangs over the Northwest Face and at the base of the wall, a group of climbers were slowly moving up the face. At about 6:30 PM, they heard Leeder fall and hit the ground. The cause of his fall is unknown.

June 2009 | Manoj Kumar, 40

A "perfect storm" of conditions moved over Half Dome the day that Kumar made his ascent. It was a Saturday, the busiest day to go up the route, and people crowded onto the Dome. A storm moved in, bringing hail, rain, and disorienting fog. On the cables, people panicked. Many stopped in their tracks, clinging to the steel and refusing to move. Others, trying to get off the lightning rod, ventured outside the cables to bypass the gridlock of fear. Kumar had reached the summit and now had to descend through this mayhem. Somewhere on the route, he slipped on the wet granite and fell 200 feet to his death. A helicopter brought two Park Rangers to assess the scene. They found Kumar, deceased from severe trauma, and they found 41 other people who needed help descending the cables. Many of the hikers were poorly dressed for the mountains, wearing cotton t-shirts. The extreme cold of the storm remained. It drained them of their strength as they clutched to metal stanchions on the exposed rock. The helicopter returned with five more Park Rangers to assist in the massive rescue effort. By 8:15 PM, they had brought the last hiker safely to the base of the route.

September 2009 | Stephen Anderson, 32

He committed suicide on the summit via a self-inflicted gunshot.

July 2008 | Levi Chitwood, 27

He committed suicide by jumping from the summit.

June 2007 | Hirofumi Nohara, 37

Nohara was climbing up the cables with friends. The cables had turned into a bottleneck of people, but Nohara was laughing and carrying on a conversation with friends, enjoying the day. They were about three-quarters of the way up the route when he slipped. His foot caught on one of the metal stanchions that hold up the cables, which spun him around and sent him sliding out of control beyond the cables. He fell 300 feet to his death.

April 2007 | Jennifer Bettles, 43

Bettles fell from the cables while descending during a storm. The cables were still in their winter position, flat against the rock. Without the stanchions in place, ascents become exponentially more difficult. The Park discourages hikers from ascending the route when the cables are down, but people do it anyway.

November 2006 | Emily Sandall, 25

Sandall was also on the cables while they were down. She slipped on wet granite and slid to her death.

September 2005 | Bela Fehrer, 35

Fehrer was an experienced Yosemite climber. He had climbed hard routes on El Capitan as well as other Valley walls. While ascending the steep Death Slabs that stretch from Mirror Lake to the bottom of the Northwest Face of Half Dome, Fehrer took a fatal fall likely due to incorrect use of his climbing equipment.

June 2004 | Donald Cochrane, 48

Cochrane was descending the steps on Sub Dome ("The Devil's Staircase") when he complained of chest pain. He fell and tumbled down 300 feet of granite slabs.

August 2001 | Vladimir Boutkovski, 24

Rock climbers camping at the base of the Northwest Face saw Boutkovski fall through the air and hit the ground. This was likely a suicide.

July 1997 | Joachim Tolksdorff, 29

Climbers found Tolksdorff's body at the base of Half Dome's Northwest Face. Tolksdorff was in poor health and left a suicide note.

August 1995 | Michael Gerde, 57

Gerde was near the top of the cables when he collapsed onto the stone. Other hikers immediately performed CPR and continued until a helicopter brought Park Rangers to the scene. He was deceased when they arrived. His death may have been the result of a heart attack.

April 1989 | John Lanham, 21

Lanham fell while climbing unroped on the cliffs below Half Dome.

November 1988 | Donald Buchanan, 86

Buchanan had worked in the park for over 50 years and had

a recent diagnosis of terminal cancer. He went on a solo hike near Half Dome and never returned. Searchers found his campsite, but they were never able to locate his body.

October 1988 | Mitchell Reno, 35

B.A.S.E. jumping is the sport of jumping from buildings, antennas, bridges (spans), and cliffs (earth) with a parachute. It is very dangerous because of the low heights involved, no backup parachute, and the narrow margin of error with extreme consequences if something does go wrong. Reno B.A.S.E. jumped from the summit of Half Dome. He waited too long to deploy his canopy and hit the ground.

August 1987 | Young Soon Lee, 35

Young Soo Lee, from South Korea, was climbing on the Northwest Face of Half Dome with his partner Choong Hyun Ji. Young was attached to an anchor while Choong traversed above him. Choong dislodged a rock, and it then struck Young in the head. Young later died from the brain injury.

July 1987 | Bones of Unknown Male

Rock climbers found a set of bones below the Northwest Face. The likely cause of death was a fall.

July 1985 | Brian Jordan, 16 and Robert Frith, 25

This incident is possibly the most famous accident to happen on Half Dome's summit. Its fame is due in part to the fact that in the aftermath two people were dead and three injured. The day owes the rest of its fame to the book Shattered Air, which is a detailed account of the fateful day.

What made this incident so unique that someone wrote an entire book about the ordeal? Their story is a classic case of poor decision-making leading to disastrous consequences; in this case, a powerful lightning storm delivered the hand of fate.

Jordan and Frith reached the summit of Half Dome in the afternoon of July 27, 1985, after making the 8-mile and almost 5,000-foot hike from Yosemite Valley. They celebrated their success with beer and food, lingering on the exposed summit for too long. They were not alone in this error; several other hikers also succumbed to the Siren's song of the summit.

An afternoon thunderstorm moved over the mountains in the distance and approached Half Dome. Despite this ominous presence looming overhead, the hikers made no haste in their descent. Perhaps they were seduced by the dopamine that flooded into their brains while approaching and then attaining the summit. Summit fever overtakes many hikers and climbers; it is a common illness. Every year people die throughout the world because they fail to heed nature's warning signs in the mountains—either continuing to the summit when they should not or staying on the summit too long.

The storm rolled overhead. The two young companions, Jordan and Frith, found shelter under a rock overhang. Some of the other hikers also hunkered under the rock. This was the same overhang where lightning struck Edward Willems in 1972.

In wilderness medicine and survival, we are taught to never take shelter in a shallow cave, overhang, or doorway during a lightning storm. The human body is an excellent conductor.

Lightning seeks the path of least resistance to the ground. A person who stands or squats underneath an overhang creates a bridge for the electricity to reach its destination, the earth.

Lightning struck the rock above the hikers. The electricity traveled through Jordan, Frith, and two others. Then another strike hit one of the other hikers.

The average lightning bolt carries over 1 billion joules of energy. That is roughly the energy a 60-watt light bulb would use running continuously for six months. Lightning can also heat the surrounding air to 50,000 degrees Fahrenheit (over 27,000 degrees Celsius); that is five times hotter than the surface of the sun!

Jordan died instantly. The electrical discharge sent Frith into severe convulsions. He started to roll towards the edge of the cliff. The other hikers tried to save him, but lightning continued to strike the mountain and impeded their efforts. Frith rolled off the summit and fell nearly 2,000 feet to his death.

Before the accident, other hikers had seen the thunderheads and decided to wait out the storm far below the summit. As this story demonstrates, others pushed forward despite the signs and then dawdled when haste was due. Sadly, this was not the first instance, nor last, of hikers venturing onto the summit when the weather was dangerous. These ill-fated hikers walked right past a sign warning of the risk of lightning on the mountain.

The lesson cannot be stressed enough: the mountains do not care. The summit of Half Dome is incredibly exposed to the weather and has no safe shelters. The steel cables are a giant lightning rod. The granite under the cables is slippery when dry and deadly when wet.

The best mountain climber is the one who returns home alive, not the one who dies reaching the summit.

August 1982 | James Tyler, 35

Tyler was an experienced parachuter who had taken up the hazardous activity of B.A.S.E. jumping. After leaping from the summit, he deployed his parachute. His parachute opened in the wrong direction and sent him into the wall. The canopy collapsed upon his impact with the wall. Without the loft of the open canopy, he fell to the rocks below.

January 1978 | Kendall West, 20

West and friends were climbing a mixed rock and snow route on the West Shoulder of Half Dome. They succeeded in reaching the top of the route, but getting down proved more difficult. They chose an unfamiliar and challenging descent. After a series of rappels, their rope became stuck. West decided to climb unroped to a nearby ledge to find a better way down. While doing this, he took a 1,000-foot fatal fall. His 13-year-old partner spent the night on a small ledge, and a helicopter rescued him in the morning.

August 1972 | Edward Willems, 19

As is often the case, a storm moved over Half Dome and brought lightning with it. Willems attempted to hide from the lightning under an overhanging rock, but he was struck and killed.

March 1968 | Larry Greene, 29 and Edwin Hermans, Jr., 24

Greene and Hermans were climbing on Half Dome when the avalanche danger was high. They either fell on the snow or got caught in an avalanche. Rescuers found the men's bodies near the bottom of an avalanche gully.

August 1956 | James Stergar, 30

Stergar left a suicide note in his car, hiked to the summit, and leaped to his death.

September 1948 | Paul Garinger, 41

Garinger was descending the cables when he stopped. Witnesses said he looked sick. Moments later he fell to his death, possibly after fainting. Garinger was the first person to die on the Cables Route. Even though the exact circumstances leading to his death are unknown, it is vital to remember that hydration and avoiding overexertion are incredibly important when at high altitudes.

June 1948 | Chalmers Groff, 19

Groff and a friend were on their weekend break from working for the National Park Service. They were climbing down the Death Slabs below Half Dome when Groff slipped on a mossy rock. He fell 70 feet, almost taking his friend with him. He died instantly.

April 1933 | Godfrey Wondrosek, 26

Wondrosek set off to climb to the summit of Half Dome. He never returned and was never found.

September 1930 | Vincent Herkomer, 23

Herkomer was a chemistry student at Berkeley. He hiked to the base of the dome and swallowed poison. Park scientists found his bones four years later.

THE CHALLENGE OF THE MOUNTAIN

For every accident that occurs on Half Dome, there are tens of thousands of successful and happy endings. Most hikers and climbers complete their ascents and leave Yosemite with fond memories of the challenges they faced and overcame in the mountains. The story of Half Dome is far richer than any book can tell, for Half Dome's story is in the hearts and minds of those who have found inspiration in this icon of nature. With the history behind us, all that remains is the status of Half Dome today.

13

Half Dome Today

The story of Half Dome today is one of continued inspiration, but also of meticulous management to retain the integrity of the Half Dome experience for all. The National Park Service works to provide access to hikers and climbers while also protecting access for future visitors and protecting the unique summit habitat.

A MOUNTAIN OF DATA

Managing a mountain involves lots of data collection. This data includes: the number of applications to hike up the cables; the number of climbers sleeping at the base; the movement patterns of overnight hikers venturing into the Half Dome area; the success of the flora and fauna eking out a life on the summit; the number of bear-human interactions in the Half Dome Trail corridor; and the value of gear destroyed by bears trying to get human food. Collecting all of this data requires scientists, researchers, rangers, databases, and lots of charts. Interpreting this information and turning it into policy requires lots of committees and public hearings.

Half Dome is a different place than it was when George Anderson first ventured onto its summit. There were no committees or permits then. However, these modern changes have

Cross-country skiers rest at Glacier Point.
Photo courtesy of the Yosemite National
Park Archives.

come for a reason: Half Dome is more popular than ever.

The 2016 Yosemite season[5] saw 51,010 applications to climb up the cables. Each application can include up to 6 people. The odds of getting a permit hovered around 12% overall. More people apply each year, and thus the chances of getting up the mountain decrease as well.

The number of overnight backpackers heading into Yosemite's wilderness grew 10% between 2015-2016 to 70,357. Many of those passed through the Half Dome corridor at one point on their trip.

THE FUTURE—KEEPING HALF DOME WILD

With more people than ever wanting to visit Yosemite and struggle their way up the Dome, whether via the cables or a steep rock climb on the Northwest Face, it is more important than ever to educate visitors about protecting the resource through simple Leave-No-Trace actions. What are these principles?

- Plan Ahead & Prepare (As evidenced in the discussion on deaths on Half Dome, all of us can benefit from more preparation.)
- Travel & Camp on Durable Surfaces (Staying on established trails reduces erosion.)
- Dispose of Waste Properly (If you pack it in, pack it out, including toilet paper. Bury feces

---

5   The latest published statistics for Half Dome as of late 2018.

200 feet from water sources.)
- Leave Artifacts and Natural Objects You Find
- Minimize Use & Impacts of Fire
- Respect Wildlife (In Yosemite, improper food storage often leads to the euthenization food-habituated of black bears, especially at the Little Yosemite Valley campground that is base camp for many Half Dome hikers.)
- Be Considerate of Other Visitors

Our actions outside of our parks also affect these lands. Global climate change from human-caused emissions will impact Half Dome. Data from the last 30 years documents a pattern of year-on-year warming. Climate models predict that temperatures in central California will be five to ten degrees Fahrenheit higher by the end of this century. This temperature change will be most profound at higher elevations. Reduced snow packs will affect water supplies for Californians because the snowfields and glaciers of the High Sierra act as huge, natural reservoirs. Higher temperatures and smaller snowpack will also threaten species such as the Lyell salamanders that live on Half Dome's summit.

The use numbers suggest that humans will play a big roll in the evolution of Half Dome's story in the coming decades. Yes, there will be the occasional rockfall that changes a climbing route. But the geological story is on a much slower timeline than the human and ecological stories. Will ever-growing numbers of people want to ascend the mountain? From a broader perspective, will the outdoors and our parks,

The Cables Route on a very busy day. Photo by Saibo.

in general, continue to grow in popularity? Will rock climbing continue its meteoric rise from a fringe activity to a mainstream sport? These questions beget more questions. As more climbers aspire to ascend Half Dome's vertical spaces, will accidents increase as well? At what point will park management be forced to limit the number of people entering the park to protect the natural resources? Will climate change herald the death of those stoic summit hermits, the salamanders?

We can even branch into the philosophical with these questions. How will technology affect people's desire to climb Half Dome and other famous peaks? Today, people can "hike" the Half Dome Trail using Google Maps. Will this decrease or increase the desire to ascend this landmark?

It is important to recall McAllister's original intent when he constructed the Cables Route. He wanted to inspire love and care for nature. Hopefully the 50,000 hikers a year attempting the route are discovering love for the wild places of this world and learning to care for them.

All of this pondering aside, we can be confident that Half Dome is not going anywhere. Cultures come and go. Governments collapse. Cities rise and fall. Half Dome, Tis-sa-ack, will be there in a thousand years and another thousand after that, standing guard, tearfully, over the sacred valley the Native Americans called Ahwahnee.

THE END.

Acknowledgments

Writing this book was a much longer process than I first anticipated, but it was an enjoyable period of time nonetheless. Further, publishing it required more than just my time. Many people graciously lent their expertise to this project, including editing, proofreading, and design.

This book would not have been possible without help from the following people. Susan Greenleaf, for her art and encouragement. Mark Scrimenti, for his excellent edits. Dannique Aalbu for always being there. Tressa Gibbard for her keen eye. Jean Redle for her attention to detail. The team of proofreaders including Sharon, Lauren, Brandon, Mary Clare, Van, Mary Ann, and Paxton. The living spirit of Yosmite, Stu Kuperstock. Beryl Knauth for her friendship and knowledge of the Golden Age. Tamara for our recent conversations about Yosemite's climbing history. Roger Derryberry and Mary Lou Long for their friendship and sharing. Nancy and the Inyo County Library. And tons of encouragement from Bud, Dan M., and Kim. Thank you to the interpreters at Yosemite National Park for doing a great job, especially in preserving Yosemite's rich cultural history, and the Yosemite Archive for preserving loads of documents and making them easily available.

Finally, there are many people not mentioned here but who helped indirectly through their friendship or mentorship over the years, including many in Yosemite. Thank you.

Bibliography

In writing *Half Dome: The History of Yosemite's Iconic Mountain*, I used many printed and digital sources. Since this is not an academic work, I decided not to clutter up the pages with reference notes. Instead, what follows is a basic, though not exhaustive, reference list.

Abrams, William Penn. Diary of October 18, 1849.

Beeler, Madison Scott. *Journal of the American Name Society*. 3 (3): 185-186, September 1955.

Browning, Peter. *Yosemite Place Names*, 1988. Great West Books. Lafayette, CA.

Diamant, Rolf. Letter from Woodstock. *Lincoln, Olmstead, and Yosemite: Time for a Closer Look*. 2014.

Ghilgieri, Michael and Charles Farabee. *Off the Wall. Death in Yosemite*, 2007.

Glazner, Allen and Greg Stock. *Geology Underfoot in Yosemite National Park*, 2010. Mountain Press Publishing Company. Missoula, MT.

Hutchings, James M. *In the Heart of the Sierras*, 1888. Pacific Press Publishing House.

Isserman, Maurice. *Continental Divide: A History of American Mountaineering*, 2017. W.W. Norton & Company.

Jones, Ray. *It Happened in Yosemite National Park: Remarkable

Events that Shaped History, 2010.

Matthes, François. *Geological History of the Yosemite Valley USGS Professional Paper 160*, 1930.

Nelson, Anton. "Half Dome, Southwest Face," *Sierra Club Bulletin* 31 (December 1946): 120-21.

Narrative of Third Sergeant Alexander M Cameron, in Bunnell, *Discovery*, 1911.

Robbins, Royal. *Fail Falling*, 2010. Giraffe Press.

Roper, Steve. *Camp 4: Recollections of a Yosemite Rock Climber*. Mountaineers Books, 1998.

Taylor, Joseph. *Pilgrims of the Vertical: Yosemite Rock Climbers and Nature at Risk*, 2010. Harvard University Press.

Taylor, Katherine Ames. *Lights and Shadows of Yosemite*, 1926.

Whitney, Josiah D. *The Yosemite Book*, 1869.

Wikipedia

Yosemite National Park. www.nps.gov/yose

Yosemite Nature Notes. www.yosemite.ca.us

Index

A

Abrams 43
Adams, Ansel 87
Ahwahneechee 39
Ahwahnee Hotel 42
Ah-wei-ya. *See also* Mirror Lake
Albright, Horace 122
Alpine House 66
American Alpine Journal 116
American Civil War 45
Ancestral Sierra Nevada 6
Anderson, George 57
Atlatl 37

B

Basket Dome 42
Batholith 6
Bierstadt, Albert 46
Black Oak 38
Broderick, Mount 86
Bunnell, Lafayette 40

C

Cables Route 71
California Geological Survey 45, 49
Cameron, Sergeant Alexander 44
Camp Curry 72
Casa Nevada 66
Cascade Cliffs 112
Cascade Range 5
Chouinard, Yvon 106

Chuck-ah 38
Clark, Galen 67
Clark, Mount 109
Conway, John 54
Corvus corax. *See also* Raven
Curry, David 72

D

Devil's Postpile 7
Devil's Staircase 74
Diving Board 87
Dutcher, Miss S.L. 64

E

Eichorn, Jules 82
Exfoliation 8

F

Farquhar, Francis 82
Feurer, Bill "Dolt" 94
Fontainebleau 58

G

Gallwas, Jerry 92
Gardiner, James 45
General Federation of Women's Clubs 122
Glacier Point 12

H

Half Dome Granodiorite 5
Harding, Warren 92

Higher Cathedral Spire 82
Honnold, Alex 101
Hutchings, James 51
Hutchings House 52
Hydromantes platycephalus. *See also* Lyell salamander

I

Igneous rock 5

K

Ka-cha-vee 39
King, Clarence 45, 49

L

Leave-No-Trace 147
Leonard, Dick 82
Liberty Cap 86
Lincoln, President 45
Little Yosemite Valley 50
Long, Dick 92
Long Valley Caldera 7
Lost Arrow Spire 89
Lost Lake 86
Lyell, Mount 19
Lyell salamander 19

M

Majestic Hotel. *See also* Ahwahnee Hotel
Mammoth Lakes 7
Mandatory, George 92
Mariposa Battalion 39, 44
Mariposa Grove 122
Marmot 23
Marmota flaviventris. *See also* Marmot

Mather, Stephen 122
McAllister, Matthew Hall 71
Mirror Lake 8, 41
Monos 39
Muir, John 34, 54, 119

N

Nangas 41
National Geographic 76
National Park Service 123
Nelson, Anton "Ax" 86
Nevada Fall 50
North American Plate 5

O

Organic Act 122

P

Paiutes 39
Peninsula Wrought Iron Works 84
Pine trees 30
Pioneer Cemetery 67
Piton 82
Pluton 5
Powell, Mark 94
Pratt, Chuck 106
Proctor, A. Phimister 68

Q

Quercus kelloggii. *See also* Black Oak

R

Raven 27
Reamer 43
Regular Northwest Face 91

Robbins, Royal 92
Roosevelt, President Theodore 122
Rowell, Galen 104
Royal Arches 42

S

Salathé, John 84
Sampson, Alden 68
San Francisco Bulletin 54
Savage, James 39
Sherrick, Mike 94
Sierra Club 71, 82
Snow, Albert 66
Snow, Emily 66
Southern Sierra Miwok 37
Spencer, Private Champion 44
Starr King, Mount 109
Steck, Allen 89

T

Tenaya, Chief 40
Tenaya Canyon 40
Tenaya Lake 40
Tenaya Peak 40
Thank God Ledge 101
Tioga Glaciation 37
Tis-sa-ack 41

U

U-ma-cha 38
Underhill, Dr. Robert 81

V

Vernal Fall 50

W

Washington Column 42
Watkins, Carlton 46
Whitney, Josiah 49
Whitney, Mount 45, 49
Wilderness Act 124
Wilson, Don 92
Wilson, Jim 92
Wilson, President Woodrow 122

Y

Yellowstone 45
Yosemite Climbing Association 35
Yosemite Facelift 35
Yosemite Grant Act 45

The Author

Joe Reidhead lives in California's Sierra Nevada mountains. An avid outdoorsman and mountain climber, Reidhead worked in Yosemite's wilderness for several years and was fortunate to live beneath Half Dome's shadow during that time. He has climbed Half Dome, El Capitan, and other granite walls in Yosemite Valley. Reidhead operates a small publishing house dedicated to mountain literature.

Made in the USA
Las Vegas, NV
23 December 2021